SCIENCE
to SIXTEEN
New edition

Stephen Pople

Formerly Head of Physics,
Withywood Comprehensive School, Bristol

Michael Williams

Head of Science,
Withywood Comprehensive School, Bristol

Oxford University Press

Oxford University Press
Walton Street
Oxford OX2 6DP

Oxford   New York   Toronto
Delhi   Bombay   Calcutta   Madras   Karachi
Petaling Jaya   Singapore   Hong Kong   Tokyo
Nairobi   Dar es Salaam   Cape Town
Melbourne   Auckland
and associated companies in
Berlin   Ibadan

ISBN 019 9142866

First published 1980
Second Edition 1987
Reprinted 1988

Printed in Great Britain by
Butler & Tanner Ltd, Frome and London

The front cover is a 'heat picture' of a man drinking a cup of tea. The rear cover photos are for use in the chapter on molecules and waves. See pages 91 and 96.

Contents

Introduction

This book is about science. Science is about asking questions. How big is the Universe? What is inside an atom? What happens to electricity after you've used it? Why do babies grow up similar to their parents, but not quite the same?

This new edition has been revised and enlarged for GCSE. It covers many of the theories that you will need to know for these examinations.

To work as a scientist, you need to ask questions. You then need to find answers! This is often done using experiments. You can also look up the answers in a reference book – if you are prepared to trust someone else's work. This reference book contains the answers to many questions that scientists have asked. This is how to use it:

Use the contents page   If you are looking for information on a large topic – the human skeleton, for example – look it up in the contents list. You will find it in *Chapter 5: Living things* on page 182. You can then turn straight to this section for the information you want. But if you can't find it on the contents page, then:

Use the index   Suppose there is something small that you want to check up on – teeth, for example. Look up 'teeth' in the index. The index gives a page number – page 183. Turn to page 183, and somewhere on it you should find what you want.

Use the questions   Asking questions and answering them is a good way of learning. To help you learn, there are questions on every section. These should help you to check that you have read and understood that section properly. At the end of each chapter, you can test yourself using the exam level questions.

Science is important   The information that scientists have gained is important. You can use this information to help understand your world, and you can use it to help understand yourself. Enjoy asking questions. Enjoy finding answers. It's part of what life is about.

Stephen Pople.
Michael Williams.

Chapter 1

Earth, air and water

What's the biggest thing you can think of?
What's the smallest?

Galaxies, stars and planets

The Sun is only one star in a huge galaxy of over one hundred billion stars. But even this vast collection of stars is not all there is in the Universe – beyond it lie many billions of other star systems.

Galaxies and stars

Each star system in the Universe is known as a *galaxy*. Our own Sun is a member of a collection of stars called *the Galaxy*.

The galaxy shown in the photo is similar to our own galaxy. Its size is enormous. On the photograph, a region the size of a pinhead might contain more than a million stars yet a journey between just two of them might take thousands of years at the speed of today's rockets. To cope with the vast size of the Galaxy, scientists measure the distances between stars in *light years*.

A light year is the distance a beam of light would travel through space in one year.

Light travels very fast – nearly 300 000 kilometres every *second* – so by Earth standards a light year is a very long distance indeed. Even the star nearest the Sun is over four light years away and appears only as a point of light. The Galaxy itself is about 100 000 light years across and the Sun lies about half way out from its centre.

The Sun

As stars go, the Sun is neither very big nor very hot, yet it has a diameter of nearly 1½ million kilometres and a surface temperature of 6000 degrees centigrade. At its centre, the temperature rises to 15 *million* degrees centigrade.

The Sun contains more hydrogen gas than anything else. A non-stop nuclear explosion deep in the Sun releases large amounts of heat as the hydrogen is slowly changed into helium gas. It is this heat that keeps the Sun shining.

The Solar System

The Sun is at the centre of a system of planets and smaller bodies which together are called the *Solar System*. The scale diagram below shows the relative distance of each planet from the Sun.

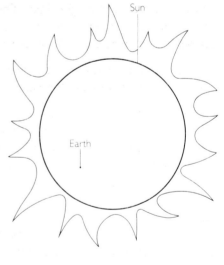

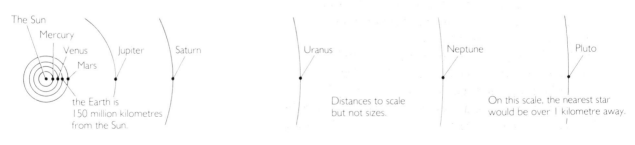

The Sun
Mercury
Venus
Jupiter
Saturn
Mars

the Earth is 150 million kilometres from the Sun.

Uranus

Distances to scale but not sizes.

Neptune

Pluto

On this scale, the nearest star would be over 1 kilometre away.

The planets move around the Sun in more or less circular paths called *orbits* – the further a planet is from the Sun the longer it takes to complete an orbit. Many planets have moons around them.

Scientists think that the planets were formed from a vast cloud of gas and dust which once encircled the Sun. All are very much smaller and cooler than the Sun. None give off their own light though they do reflect some of the sunlight which strikes them – it is difficult to tell at a glance whether a point of light in the night sky is a star or a planet.

Mercury is not much larger than our own Moon. It has a crater-pitted surface and conditions are hot and hostile.

Venus has a thick atmosphere rich in carbon dioxide and sulphuric acid gases. The surface of the planet is very hot.

The Earth is the only planet in the Solar System known to have the conditions needed to support life.

Mars is often called the Red Planet because of its rust coloured surface. It has a thin atmosphere mostly of carbon dioxide gas and polar caps of frozen carbon dioxide and water.

The asteroids are many thousands of minor planets which orbit the Sun. The largest, Ceres, is less than 700 kilometres across.

Jupiter is more massive than all the other planets put together. It is made up very largely of hydrogen.

Saturn is encircled by bright rings. These are made up of billions of particles of rock and dust, each like a tiny moon. These particles may be the remains from a moon that broke up.

Uranus, Neptune and Pluto are the coldest of the planets because of their great distance from the Sun.

Many other bits of rock, dust, ice and even gas orbit the Sun. *Meteors* are tiny specks of rock which flare up if they hit the Earth's atmosphere. Larger rocks strike planets now and again – these are called *meteorites*. *Comets* are collections of ice, gas and dust, which reflect light from the Sun.

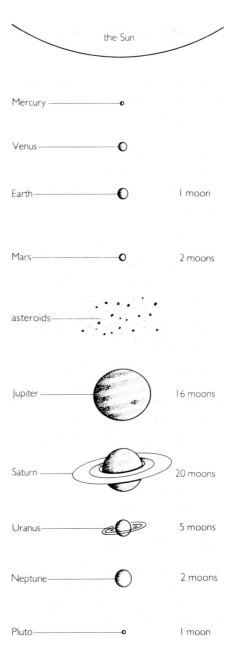

[sizes to scale but not distances]

Questions
1. What is (a) a galaxy (b) the Galaxy (c) a *light year?*
2. How far away is the star nearest the Sun?
3. What is the temperature on the surface of the Sun?
4. What is the most common substance in the Sun?
5. How far is the Earth from the Sun?
6. Which is the largest planet in the Solar System?
7. Which planet is sometimes called the Red Planet?
8. Which planet has most moons in orbit around it?
9. Which planet in the Solar System is furthest from the Sun?
10. Which planet takes longest to complete one orbit of the Sun?

The Earth

The Earth was formed 4500 million years ago from part of the cloud of dust and gas around the Sun. As the dust and gas started to cling together a ball of hot rock formed. It's not so hot today – there's water on the surface and air around it, but there's plenty happening under the surface.

Zones of the Earth

The deepest holes that humans have drilled are no more than pin-pricks on the surface of the Earth. The Earth is about 12 600 km across. By studying the waves sent out by earthquakes, scientists now believe that there are three zones inside the Earth.

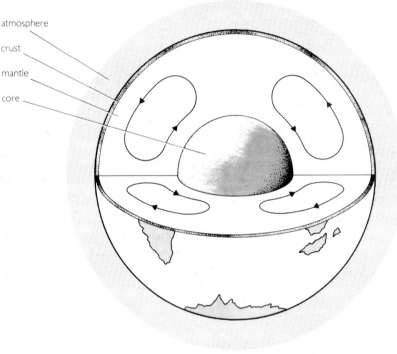

atmosphere

crust

mantle

core

1. Core The central *core* is very dense, extremely hot rock and is under great pressure. It is partly liquid and consists of the same kind of material as meteorites.

2. Mantle Surrounding the core is a thick layer called the *mantle*. The rocks in this layer are also heavy but less dense than the core. The pressure is lower nearer the outside and the rocks are cooler, although still hot enough for some to be liquid. Convection currents (explained on page 84) cause slow movements of the rocks within the mantle. In places, rock from the mantle has pushed up through the outer layer and is exposed on the surface.

3. Crust The thin outer layer is made of rock less dense and much cooler than the mantle beneath it. Under the oceans it is only 6 km thick and is no more than 70 km thick under the highest mountain peaks.

The crust is something like the shell of a cracked egg; it is made of a number of large sections called *plates*. These plates are moved very very slowly by the currents in the mantle below. Where the plates rub against each other, pressure builds up. From time to time cracks appear on the surface (*faults*) and violent shaking movements occur (*earthquakes*). Mountains formed in past ages when plates were pushed together and the rocks at the edges were forced upwards.

The results of an earthquake.

8

Types of rock in the crust

There are three different kinds of rock in the crust.

1. Igneous rock   The deepest layer of the crust, next to the mantle, is rock called *basalt*. This is the rock under the oceans. On top of the basalt layer are huge rafts of *granite* which make up the continents. These igneous rocks make up 90% of the crust but they do not show on the surface very much except where there have been volcanoes. *Volcanoes* occur when hot liquid rock, formed in pockets where plates meet, forces its way through cracks to the surface and bursts out as dust, gas, and burning liquid *(lava)*.

Section through a lump of granite.

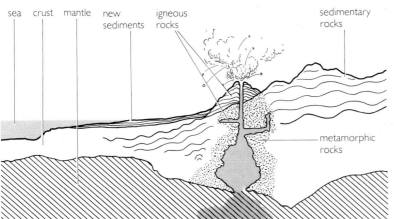

sea   crust   mantle   new sediments   igneous rocks   sedimentary rocks   metamorphic rocks

2. Sedimentary rock   Wind, rain and frost break down the rocks on the Earth's surface. The small bits are washed or blown away and deposited in layers, usually where rivers meet the sea. As the layers build up, the lower ones become compressed and hardened into sedimentary rocks. Mountain-building movements cause the hardened layers *(strata)* to tilt and fold. This process has been repeated again and again for over 3000 million years. Most of the Earth's surface is covered with a thin layer of sedimentary rock. *Limestone* and *sandstone* are common examples.

Mount Vesuvius erupting.

3. Metamorphic rock   Sedimentary and igneous rocks can be changed by heat and pressure into different, often harder, forms. Heat from a volcano sometimes causes surrounding limestone rocks to change into *marble*. Compressed mud can be changed by pressure into *slate*.

Questions

1. How long ago did the Earth form?
2. What is the depth of the centre of the Earth from the surface?
3. How thick is the crust under (a) oceans, (b) high mountains?
4. What are the large sections of the crust called?
5. What causes earthquakes? How were mountains formed?
6. What kind of rock comes from volcanoes?
7. How do sedimentary rocks form? Why are they often folded?
8. Name two igneous, two sedimentary and two metamorphic rocks.

Folded rocks exposed in a cliff.

Minerals, elements and atoms

Substances which can be dug out of the Earth's crust and which are valuable to us are called minerals. But what are minerals made of?

Minerals

The term mineral is rather vague. It usually means substances which are found in a fairly pure state. Diamonds, quartz and alum crystals are examples.

An uncut diamond.

Alum crystals of different shapes.

Pure minerals are formed from rocks by the actions of heat, pressure or water. Some are extremely valuable (gold and diamonds), and others are important because of the substances which can be made from them. Rock salt, for instance, provides us with chlorine gas and household bleach.

Not all minerals are pure solids however. Coal, crude oil and natural gas are some of the impure materials in the Earth which are important minerals.

Quartz crystals.

Elements

Many minerals can also be broken down into even simpler substances. The simplest of all are *elements*. They cannot be broken down further.

About 90 elements–joined together in lots of different ways–make up all the materials of the Earth. Oxygen is the most abundant element. It is present in air, water and nearly all rocks, making up about half the total weight. The next most common element is silicon. The diagram shows the amounts of the nine most common elements in the Earth's crust.

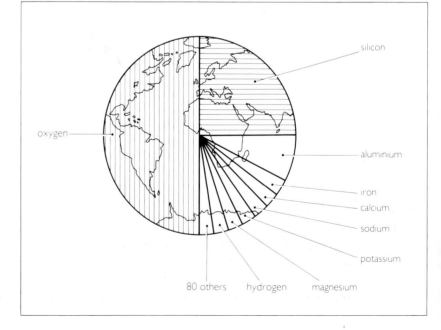

The two kinds of elements

Elements may be grouped into two kinds: metals and non-metals.

Metals   These are hard, shiny solids which only melt on strong heating.

They can be bent, drawn out into wires or beaten into shapes. They are good conductors of heat and electricity.

Non-metals   These are either gases or solids which melt easily. The solids are usually powdery or brittle. They do not bend. They are poor conductors of heat and electricity.

Some elements do not show all the properties typical of their kind. Carbon, for example, is a non-metal but, in the form of diamond, it is the hardest substance known.

Atoms

An iron nail appears to be solid but it can be cut into smaller pieces. The smallest pieces which can exist by themselves are called *atoms*. All elements are composed of atoms. The atoms of an element are all alike but they are different from the atoms of any other element.

Atoms are so incredibly tiny that it is difficult to picture just how small they are. A penny, for instance, contains about 34 thousand, million, million million atoms!

Atoms of elements are often shown as *symbols*. The symbol is either the first letter of the name (sometimes the Latin name) or the first letter followed by a small letter. Thus, the atom of carbon is C, calcium is Ca, copper is Cu (from cuprum). Other symbols are shown below.

Iron in the pylon and aluminium in the cables – both typical metals.

Metals			
aluminium	Al	magnesium	Mg
calcium	Ca	potassium	K
copper	Cu	silver	Ag
gold	Au	sodium	Na
iron	Fe	tin	Sn
lead	Pb	zinc	Zn

Non-metals			
bromine	Br	iodine	I
carbon	C	nitrogen	N
chlorine	Cl	oxygen	O
fluorine	F	phosphorus	P
helium	He	silicon	Si
hydrogen	H	sulphur	S

Questions

1. What kind of substances are called minerals?
2. How are minerals formed? Name four minerals.
3. What are elements? How many are there in the Earth?
4. What are the two most abundant elements?
5. Give three properties each of metals and non-metals.
6. What are the smallest possible pieces of elements called?
7. About how many atoms are there in one penny?
8. Give the symbols for sodium, calcium, iron, oxygen, carbon and copper.

Inside atoms

The idea of atoms being the smallest possible pieces of an element has been around for thousands of years. It is now known that the 90 or so different atoms are themselves made up of even smaller parts.

Sub-atomic particles

Of the many tiny particles which make up atoms, only three, *protons*, *electrons* and *neutrons*, are needed to explain how atoms differ from each other.

The proton p, and the electron e   Protons are packed together as a central core or nucleus with electrons orbiting round, somewhat like planets round the Sun. Electron paths are constantly changing but they are usually drawn as circles. Protons are nearly 2000 times heavier than electrons.

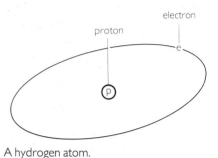

A hydrogen atom.

The proton and the electron have a mysterious property called *charge*. Charge exerts a force of attraction between the particles and holds the atom together. Protons are said to have positive (+) charge and electrons to have negative (−) charge. Atoms always have equal numbers of protons and electrons so that the charges are balanced. The simplest atom, hydrogen, has one proton and one electron. The next atom, helium, has two of each, whilst the heaviest naturally-occurring atom, uranium, has 92 of each particle.

The neutron n   This is about as heavy as the proton but is not charged. Most atoms have neutrons packed together with the protons in their nuclei. The helium atom, for example, has two neutrons in addition to two protons in its nucleus.

Proton number and nucleon number   The proton number of an element (sometimes known as the *atomic number*), tells you the number of protons in each atom of the element. For example, helium has a proton number of two. Other elements have different proton numbers.

Sir James Chadwick, who discovered the neutron in 1932.

The nucleon number of an element (sometimes known as the *mass number*), tells you the total number of protons and neutrons in an atom. Helium, with two protons and two neutrons, has a nucleon number of four.

The helium atom is sometimes shown as $_{2}^{4}\text{He}$ ^{4} ------ mass number $_2$ ------ atomic number

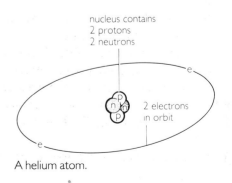

nucleus contains
2 protons
2 neutrons

2 electrons
in orbit

A helium atom.

Isotopes

All atoms of a particular element have the same proton number. They have the same number of protons in the nucleus and electrons in orbit. The number of neutrons is not fixed, however, and different forms of the same element can exist. Every lithium atom, for instance, has three protons, but some also have three neutrons (nucleon number = 6), whilst others have four neutrons (nucleon number = 7). These two forms of the element are called *isotopes*. A great many elements are mixtures of isotopes.

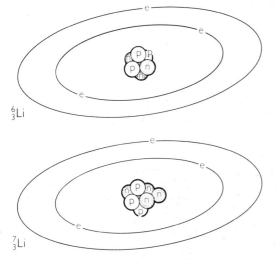

first shell contains 2 electrons

second shell contains 8 electrons

third shell contains 1 electron

A sodium atom.

Electron shells

The electrons in an atom orbit the nucleus in clearly defined regions called *energy levels* or *shells*. The first shell is close to the nucleus and has room for only two electrons. The second shell can contain up to eight electrons. There can be as many as seven occupied shells of electrons. Lithium has three electrons: two in the first shell and one in the second. Sodium has 11 electrons: two in the first shell, eight in the second shell and one in the third. The shells nearest to the nucleus fill up first.

The way an element behaves largely depends upon the electrons in the outermost shell. These outer electrons are the ones most commonly involved when atoms join up with each other. There are some atoms which do not join up with others: these atoms usually have their outer shells filled with eight electrons.

Questions

1. Name the particles present in the nucleus of an atom.
2. What is it that holds an atom together?
3. What is the proton number of an atom?
4. Which element has a proton number of 92?
5. What is nucleon number and what does it show?
6. What are isotopes? Give an example.
7. What are electron shells? How many electrons will the 2nd shell hold?
8. Which electrons are involved when atoms join together?

Niels Bohr was the Danish scientist who first worked out the idea of electron shells.

13

Atoms joining together

The world is not made up of separate atoms floating about – atoms stick together. Sometimes just a few join up: sometimes thousands are linked together in some way. When two or more atoms join together as a single unit, the new particle is called a molecule.

Molecules of elements

Many elements are made up of molecules which contain two or more identical atoms linked together. Air consists mainly of the elements nitrogen and oxygen: both gases have molecules made up of two atoms strongly bound together. The yellow, powdery element sulphur consists of molecules having eight atoms each.

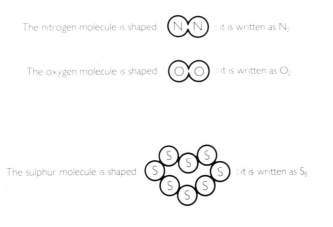

The nitrogen molecule is shaped : it is written as N_2

The oxygen molecule is shaped : it is written as O_2

The sulphur molecule is shaped : it is written as S_8

Molecules of compounds

Molecules made from two or more different atoms linked together form *compounds*. Water, perhaps the most common compound of all, has simple molecules in which two atoms of hydrogen are joined to one atom of oxygen. The molecule is represented by the formula H_2O.

The *formula* of a compound shows exactly the numbers of all atoms present in the molecule. Note that the formula for water is H_2O and not H_2O_1. If only one atom of any kind is present the symbol alone is shown without the figure $_1$. The formulae of some common compounds are shown in the following table.

A water molecule model.

Compound	Molecule	Formula
carbon dioxide		CO_2
ammonia		NH_3
methane		CH_4
sulphuric acid		H_2SO_4
hydrogen chloride		HCl

Equations

When substances combine together to form different materials, a *reaction* or *chemical change* has taken place. The burning of a match is a good example of a chemical reaction. *Equations* are a shorthand way of describing chemical reactions, using symbols and formulae to show how atoms and molecules rearrange themselves.

When coke (a form of carbon) burns in air (containing oxygen) to form carbon dioxide the equation would be written:

carbon + oxygen → carbon dioxide

$$C + O_2 \rightarrow CO_2$$

The equation shows that one atom of carbon combines with one molecule of oxygen to give the molecule of carbon dioxide.

The equation for the burning of natural gas (methane) is:

methane + oxygen → carbon dioxide + water

$$CH_4 + 2O_2 \rightarrow CO_2 + 2H_2O$$

The equation shows that one molecule of methane combines with two molecules of oxygen to give one molecule of carbon dioxide and two molecules of water.

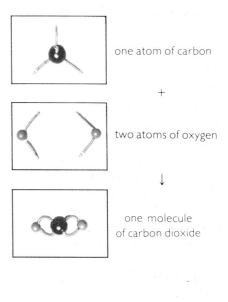

one atom of carbon

+

two atoms of oxygen

↓

one molecule of carbon dioxide

Covalent bonding

The way in which atoms join to form molecules is related to the number of electrons in their outermost shells.

A full shell of eight electrons is a stable state. Atoms will share pairs of electrons between them if, by doing so, they can achieve full outer shells. Thus two chlorine atoms, both having seven electrons in their outer shells, will share two electrons, one from each atom, to make up shells of eight.

The shared pair of electrons acts as a 'glue' sticking the two atoms together. It is called the *covalent bond* and is sometimes drawn as a line between atoms:

Cl—Cl

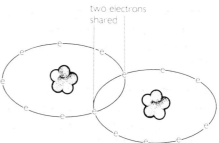

two electrons shared

Atoms of chlorine showing outer shells of electrons only.

Questions
1. What is a molecule? Give two examples of molecules of elements.
2. What are compounds composed of? What does the formula of a compound show? Give the formula for sulphuric acid.
3. What is meant by a 'chemical reaction'?
4. What do equations represent?
5. Write the equation for the burning of natural gas.
6. What part of an atom is involved in bond formation?
7. Show how two chlorine atoms combine to form a chlorine molecule.
8. What is the covalent bond?

Ions joining together

Chloroform and common salt are both compounds containing chlorine atoms. Why are they so very different?

Ionic compounds

When sodium is heated in chlorine, the elements combine to form the compound sodium chloride. It is the electrons in the outer shells of the atoms that are involved in the bonding. In this case they do not form the 'electron glue' that sticks covalent compounds together as explained in the previous section. Instead an electron is transferred from the sodium atom to the chlorine atom so that each atom has a full shell of eight electrons on the outside.

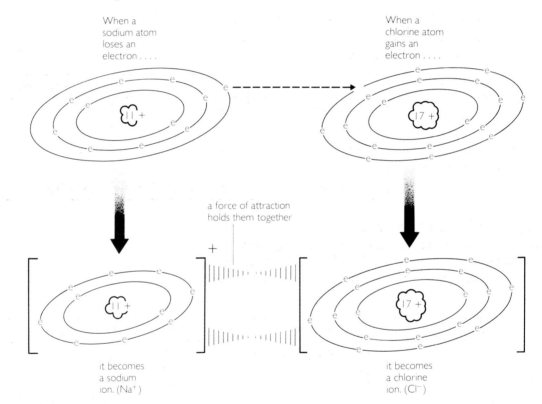

When a sodium atom loses an electron . . . .

When a chlorine atom gains an electron . . . .

a force of attraction holds them together

it becomes a sodium ion. (Na^+)

it becomes a chlorine ion. (Cl^-)

Ions Sodium chloride does not consist of molecules but of *ions*. Ions are atoms which have either lost or gained electrons. The sodium ion Na^+ is positively charged because it has lost a negative electron. When the chlorine atom accepts an extra negative electron it becomes the negative ion Cl^-. The compound is held together by force of attraction between the oppositely charged ions. This is called *ionic bonding*.

Other atoms may gain or lose two electrons and some three electrons. Complex ions are formed when groups of atoms lose or gain extra electrons.

Positive ions (cations)	
calcium	Ca^{2+}
copper	Cu^{2+}
iron	Fe^{3+}
ammonium	$(NH_4)^+$

Negative ions (anions)	
oxide	O^{2-}
sulphate	$(SO_4)^{2-}$
carbonate	$(CO_3)^{2-}$
nitrate	$(NO_3)^-$

Structure and formulae of ionic compounds

Ionic compounds have rigid structures something like the framework of a large building. Positive and negative ions are placed alternately. The structure is held together very tightly because all the positive ions attract all the negative ions close to them. The ions fit together as closely as possible.

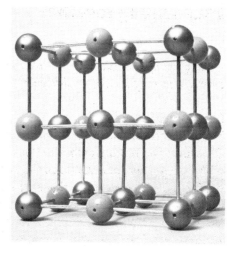

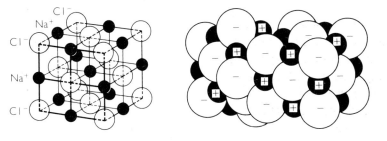

Formula The formula of an ionic compound shows the ratio of the ions present. It is the smallest number of each ion which will provide a balance of positive and negative charges.

The formula represents the smallest part of the compound which can take part in a chemical change. It may be used in equations in the same way as the formula of a covalent molecule.

Compound	Ions	Formula
sodium chloride	$(Na)^+$ $^-(Cl)$	$NaCl$
copper sulphate	$(Cu)^{++}$ $^{--}(SO_4)$	$CuSO_4$
calcium nitrate	Ca^{++} $^-(NO_3)$ $^-(NO_3)$	$Ca(NO_3)_2$

General properties of compounds

Chloroform and common salt have quite different properties because of the way they are bonded. The differences between them are typical of the differences between covalent compounds and ionic compounds.

Property	Covalent compounds	Ionic compounds
melting point	low – often gases or liquids	high – solids
smell	yes	no
dissolve in water	no	yes
conduct electricity	no	yes – when melted or dissolved

Questions

1. What is formed when sodium is heated in chlorine? What happens to the electrons in the outer shells of the atoms?
2. Give the two ions present in sodium chloride.
3. What is ionic bonding? Give the formula of the sulphate ion.
4. Describe the structure of an ionic compound.
5. What is the formula of an ionic compound? What does it show?
6. Write the formula for (a) copper sulphate, (b) sodium carbonate.
7. State two general differences between covalent and ionic compounds.

Reactions

When substances join together, break up, or form new materials chemical reactions are said to have occurred. Reactions can happen quickly or slowly and they all give out or take in heat.

zinc dust
+
sulphur powder

Types of reactions

Joining reactions (synthesis)   Magnesium burns very brightly in oxygen: the two elements link up to form magnesium oxide ash. This is an example of a synthesis in which two substances (sometimes more) join to make a single *new* substance.

magnesium + oxygen → magnesium oxide
$$2Mg \; + \; O_2 \; \rightarrow \; 2MgO$$

When zinc dust and powdered sulphur are heated together, they suddenly combine with a bright flash to form a mushroom cloud of zinc sulphide.

zinc + sulphur → zinc sulphide
$$Zn + \; S \; \rightarrow \; ZnS$$

Breaking up reactions (decomposition)   A warm solution of glucose, with a little yeast added, splits up to form the simpler substances ethanol and carbon dioxide.

glucose $\xrightarrow{yeast}$ ethanol + carbon dioxide
$$C_6H_{12}O_6 \longrightarrow 2C_2H_5OH + \; 2CO_2$$

Calcium carbonate is decomposed by strong heat (see page 48).

Displacement reactions   When an iron nail is held in a solution of copper sulphate, a little of the iron dissolves. It displaces some of the copper from the solution and this copper is deposited on the surface of the nail as a red-brown coating.

iron + copper sulphate → copper + iron sulphate
$$Fe + \; CuSO_4 \; \rightarrow \; Cu \; + \; FeSO_4$$

The displacement of hydrogen from acids by metals is shown on page 44.

Redox reactions   When hydrogen gas is passed over heated copper oxide, the hydrogen joins up with the oxygen making water and leaving pure copper.

copper oxide + hydrogen → water + copper
$$CuO \; + \; H_2 \; \rightarrow H_2O + \; Cu$$

The hydrogen has gained oxygen and is said to have been *oxidised*. The copper oxide has lost oxygen and is said to have been *reduced*. Reduction and oxidation are two parts of the single process–redox. Oxidising agents are usually substances which contain oxygen and give some or all of it to a reducing agent. The oxygen does not always add on directly to the reducing agent: sometimes it removes hydrogen (as water) and sometimes it removes electrons.

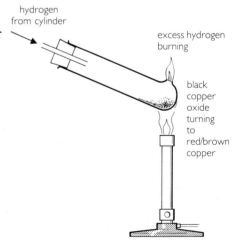

hydrogen
from cylinder

excess hydrogen
burning

black
copper
oxide
turning
to
red/brown
copper

Heat change in reactions

Put a light to natural gas (methane) and it burns, combining with oxygen in the air to give carbon dioxide and water. There are two stages to this process. First, the methane and oxygen molecules split up into single atoms. Energy is needed to break the bonds holding the molecules together so heat must be supplied (as a spark or a match) to start the reaction. Secondly, the single atoms join up making the new bonds of the carbon dioxide and water molecules. Heat energy is released in this new bond making.

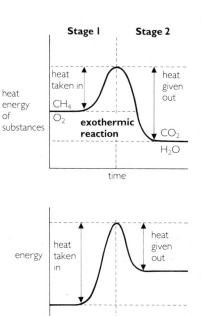

methane + oxygen →   single atoms   → carbon dioxide + water

$$CH_4 \;+\; 2O_2 \;\rightarrow\; C+H+H+H+H \;\rightarrow\; CO_2 \;+\; 2H_2O$$
$$O+O+O+O$$

        stage I   **stage 2**

In this reaction much more heat is given out in **stage 2** than is taken in during **stage 1**. Such a reaction gets hot and is said to be *exothermic*. Some reactions take in more heat in **stage 1** than is given out in **stage 2**. These become colder and are said to be *endothermic*.

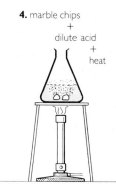

Speed of reactions

Put two marble chips into dilute hydrochloric acid and they fizz. They dissolve away slowly and carbon dioxide is given off (see page 32). The reaction time can be taken by starting a timer when the chips are added and stopping it when the marble has all dissolved. Repeat the process using crushed marble and the reaction time is much less. You can also speed up the reaction if you use hot acid or concentrated acid.

Some reactions are speeded up by light energy (as in photosynthesis), and some between gases are affected by increasing the pressure. Many are speeded up by adding small amounts of substances (*catalysts*) which are not used up in the reaction.

Questions

1. Explain what is meant by 'synthesis' and give one example.
2. What is meant by 'decomposition'? Name the products and write the equation for the decomposition of calcium carbonate.
3. What happens to an iron nail in copper sulphate solution?
4. What is an oxidising agent?
5. What are (a) exothermic and (b) endothermic reactions?
6. State four possible ways of speeding up a chemical reaction.
7. What is a catalyst? For what reaction is yeast used as a catalyst?

Water

Life on this planet began in the water that covers much of its surface. Water and air together provide the conditions which support life as we know it. This section and the following three are concerned with that most essential substance – water.

Water in living things

Plants make their food and their own bodies from water and air. More than nine-tenths of a lettuce, for instance, is simply water. There is much more water than any other substance in the human body. It is present both combined in the materials that make up the cells and in the juices which carry out the body functions.

The water cycle

The sea is the source of all water in the world. The Sun heats the sea and some of the water becomes vapour. The vapour collects as clouds and when these are blown over high ground they release the water as rain. Rain seeps through the ground and runs as rivers back to the sea. Plants absorb water from the ground and return a lot of it to the air. Rain water is collected in reservoirs, supplied to houses and industries and returned after treatment to the sea. So water circulates over and over again.

80% water

67% water

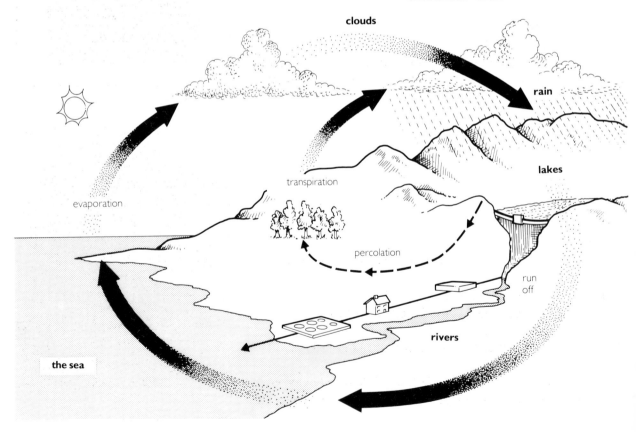

clouds

rain

lakes

transpiration

evaporation

percolation

run off

rivers

the sea

Water as a dissolving agent

One of the most important properties of water is its ability to dissolve things. If you shake a sugar cube with water, the sugar vanishes. It is still there of course – you can taste it – but it has *dissolved:* it has mixed in with the water. The millions of sugar molecules which made up the cube are now evenly spread among the billions of water molecules. The mixture of sugar and water is called a *solution*. Substances which dissolve in water are said to be *soluble*. All sorts of things are soluble in water to some extent.

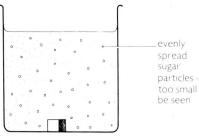

evenly spread sugar particles – too small to be seen

The sea   This is a gigantic solution of nearly all the substances found on the Earth. Rain water dissolves compounds from the rocks and soil and carries them into the sea. Common salt (sodium chloride) is the most abundant dissolved solid and makes up 2.7% of the sea. The percentage varies because the sea evaporates at different rates at different times and places.

Dehydration

Potatoes are nearly eight tenths water. When they are carefully warmed they lose their water and a solid is left. This is *dehydrated* potato. It will keep for a very long time because the bacteria and moulds which cause food to go bad cannot work without water. Powdered dried potato can be made into mashed potato by simply adding water. Other foods commonly dehydrated are: milk (for babies), peas and fruit; whilst hay is dried grass.

Tests for water

Pure water freezes at 0 °C and boils at 100 °C.
Blue copper sulphate crystals turn white when they are dehydrated by heat. The dehydrated copper sulphate turns blue again when water is added.

Dried potato.

Questions

1. How much water is there in (a) a lettuce, and (b) the human body?
2. Explain what happens in the water cycle.
3. What happens to sugar when it dissolves in water?
4. Explain what is meant by the term 'soluble'.
5. What is the main dissolved substance in sea water? How did it get there?
6. What is meant by 'dehydration'? Name 3 dehydrated foods.
7. Give the boiling point and freezing point of water and one other test for water.

Another form of dried food.

Solutions

Blood, saliva and urine are just three of the many solutions produced by the body. All the processes of living depend upon the ability of water to carry substances round the body as various kinds of solutions.

Aqueous solutions

A uniform mixture of the molecules of two different substances is a *solution*. The substance forming the bulk of the solution is called the *solvent*, and the dissolved substance is the *solute*. Solvents are usually liquids. The most common solvent is water and its solutions are given the name *aqueous*.

Solubility   Usually there is a limit to how much solute will dissolve in a fixed amount of solvent. When no more solute will dissolve, the solution is *saturated*. The *solubility* of a substance in water is the mass in grammes which will dissolve in 100 g of water to produce a saturated solution. Hot water often dissolves more of solids than cold water: the temperature has to be stated when solubility figures are quoted.

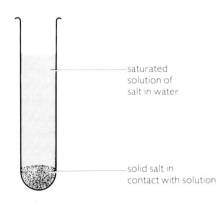

saturated solution of salt in water

solid salt in contact with solution

Solutions of gases   Most gases will dissolve in water although their solubilities are much lower than those of solids. Oxygen, for instance, dissolves very little, but there is enough of it in water for fish to breathe using their gills.

Solubilities – grams of solute in 100 g of water at 20 °C			
Solids		**Gases**	
sodium chloride (salt)	36.0	nitrogen	0.002
sodium carbonate (soda)	21.5	oxygen	0.004
ammonium sulphate	71.0	carbon dioxide	0.169
calcium sulphate	0.2	hydrogen sulphide	0.385

Suspensions and colloids

The particles of a solute in a solution are about the same size as the molecules of the solvent. Particles many times larger than the solvent molecules will not dissolve to form solutions: they are either *suspensions* or *colloids*.

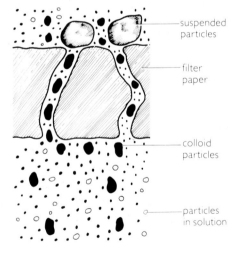

suspended particles

filter paper

colloid particles

particles in solution

Suspensions   These consist of particles large enough to be seen, mixed with water. The particles are not dissolved and will settle to the bottom if the suspension is allowed to stand. Powdered calcium carbonate shaken with water forms a milky liquid. This can be separated quickly by filtering. The large particles of the suspended solid will not pass through the fine holes in filter paper.

Colloids   A little concentrated iron (III) chloride solution poured into a beaker of hot water produces a brown, cloudy liquid. This will not clear on standing nor by filtering. It is a colloid. Its particles are bigger than those of a solution but smaller than those of a suspension. A beam of light from a strong torch will show up the particles.

Colloidal solutions

Many natural substances are colloidal. Blood contains colloidal particles as well as substances in solution. Common colloids are glue, butter, potter's clay, ink, smoke and fog.

Pastes These are concentrated suspensions of colloidal solids in liquid solvents. Paints, putty, dough and toothpaste are examples.

Jellies These are colloidal solutions even though they are semi-solid. They melt easily when warmed.

Emulsions Cooking oil, shaken with water and a few drops of detergent, produces a milky liquid consisting of tiny oil drops suspended in water. This is an emulsion and it does not separate. Common emulsions are milk, salad cream and ointments.

Some common colloids.

Other kinds of solution

Water is by no means the only liquid solvent. Many liquids have good solvent properties for particular solutes. Oil paint dissolves in white spirit, grease dissolves in petrol and nail varnish is dissolved by acetone.

All the examples of solutions considered so far have been of solid, liquid or gas solutes in liquid solvents. However, the solvent can also be a solid or a gas. Solder does not seem to be a solution, but it is actually solid tin dissolved in solid lead. Other examples of unusual solutions are shown in the photographs.

Brass is a solution of zinc metal in copper.

Smoke is a colloidal solution of solid in air.

Questions

1. Explain the terms 'solution', 'solvent' and 'solute'.
2. What is the commonest solvent? Name three other solvents.
3. When is a solution said to be saturated?
4. Explain what is meant by 'solubility'. How does temperature affect it? Give the solubilities of salt and oxygen.
5. How is a colloidal solution recognised? What size are its particles?
6. Give examples of four different kinds of colloidal solutions.
7. Name a good solvent for grease.
8. Give two examples of solutions with non-liquid solvents.

Air is a solution of oxygen in nitrogen.

Separating mixtures

Good cooks sieve flour to remove large lumps; keen gardeners sift stones from the soil of seed beds; in limestone quarries the crushed rock is graded into piles of the same-sized chippings. In the home and in industry, mixtures (including solutions) often need to be separated into their component parts.

Filtering

A *filter* is a very fine sieve made from interlocking fibres or from solid particles packed closely together. Gauze pads will filter dust and liquid sprays from air. Large, suspended solids can be removed from water by fine filter paper or by a pad made of fine powder. The smaller particles of a dissolved solid will, however, pass through a filter paper.

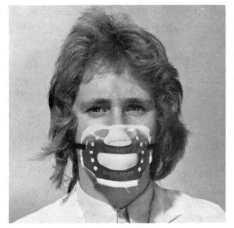

A gauze mask will filter dust from air.

Separating salt from sand   Rock salt can be purified by dissolving it in water and filtering off the insoluble sand and mud. The salt solution which passes through the fine pores of the filter paper can be evaporated to give purified white salt.

Distilling

Distilling separates and recovers the solvent from a solution. Distilled water, needed for car batteries, is obtained by boiling water and condensing the vapour. All the dissolved impurities are left behind and the distillate is absolutely pure water. Sea water can be made fit for drinking by distillation.

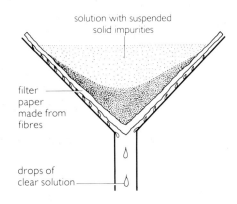

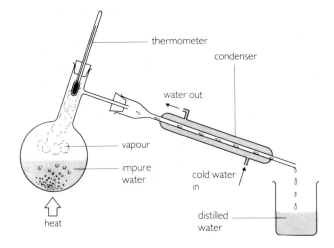

Fractional distillation   This separates mixtures of liquids. Whisky is distilled by boiling a fermented mixture of barley, water, yeast and some ethanol. The water and ethanol become gases. Much of the water condenses in the long necks of the boiling vessels and runs back. Some of the water and all of the ethanol vapour pass into the condensers to form a spirit much richer in ethanol.

Crystallising

If a little sodium chloride is stirred with water in a beaker, the salt will dissolve easily to produce a *dilute* solution. It tastes only slightly of salt. Left standing in a warm place for several days, the water gradually *evaporates* – it passes into the air as water vapour. The remaining salt solution becomes more and more *concentrated*. Eventually it becomes saturated and any further evaporation causes solid salt to be deposited on the bottom of the beaker. This is *crystallisation*. The slower the rate of evaporation, the larger are the crystals that form.

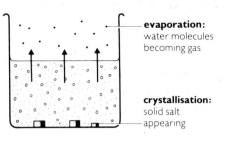

evaporation:
water molecules becoming gas

crystallisation:
solid salt appearing

Chromatography

Mixtures of solids can be separated by carrying away the lighter particles with water, oil or air. Water is used to wash sand and dirt away from gold dust and from tin ore. A similar idea is used in *chromatography* to separate mixtures of dissolved substances. Dyes in a spot of black ink on a filter paper can be separated by dripping water slowly on to the blot. As the water spreads out it carries the dyes with it. The dyes are slowed down by the filter paper by different amounts and separate into bands of colour. The dyes in different coloured felt tip pens can be compared by marking a series of dots on blotting paper which is then dipped in a bowl of water. Water soaks up into the paper drawing the dyes up to different heights. The same dyes in different inks will rise to the same heights. Non-coloured substances can also be identified by this method which is used a great deal in medical work.

Panning for gold. If there is any, the gold stays whilst other less dense substances get washed away.

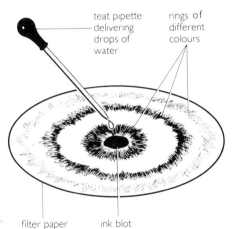

teat pipette delivering drops of water

rings of different colours

filter paper   ink blot

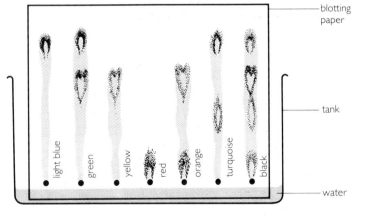

blotting paper

tank

water

light blue   green   yellow   red   orange   turquoise   black

Questions

1. What is a filter? Give examples of two different kinds.
2. Explain how rock salt can be purified.
3. Explain how distilled water is made.
4. What is 'fractional distillation'? Give an example.
5. Explain the terms 'evaporation' and 'crystallisation'.
6. What is meant by 'chromatography'? How can you show that black ink contains a number of different dyes?
7. In what kind of work is chromatography used?

Water supplies

Because our bodies are two-thirds water, we need to take in about two litres a day to remain healthy. We use many times this amount in washing, cooking and flushing the lavatory. Include the water used by industry and each of us uses about 300 litres a day.

Collecting water

There is a huge amount of water in the oceans – but it is not pure. Fresh water can be obtained easily, although expensively, by the large scale distillation of the sea. Britain is lucky to be supplied with more than enough water, distilled for nothing by the Sun. All we have to do is to tap the various sources that collect rain and make sure that the water is fit for drinking.

Mountain reservoirs   Much of our water is collected by building dams across streams in upland areas. The water is piped direct to cities or used to control the flow of rivers.

Rivers   Some towns get their water from the nearest point on a big river. This water needs careful treatment because it is usually muddy and sometimes contains harmful wastes.

Wells   Natural underground reservoirs often fill with water seeping through the ground. Well water was once very common, and still provides about a quarter of England's water supply.

Collecting water at Claerwen dam, in Wales.

Purifying water

To make water safe for drinking it goes through four processes.
1. It is passed through coarse and fine *strainers* which remove floating objects, suspended solids and organisms such as algae.

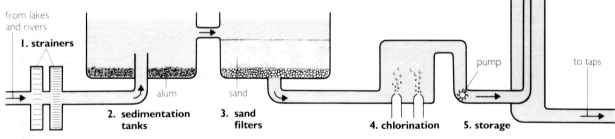

2. If the water is muddy, a *settling agent* (usually alum) is added. The colloidal particles clog together to form a spongy mass (floc) which settles out in sedimentation tanks.
3. The water is filtered through sand beds. This produces clear, clean water.
4. The clear water still contains bacteria and other micro-organisms. These are destroyed by adding carefully measured amounts of *sterilising agents* which are harmless to humans in low concentrations. Ozone gas is the best, but very costly: chlorine gas is more commonly used.

Treating used water

Once water has been used it becomes *sewage* – water mixed with organic wastes (human excrement and unused food), grease, soap and dirt. It flows away through drains into main sewers and on to *sewage* works.

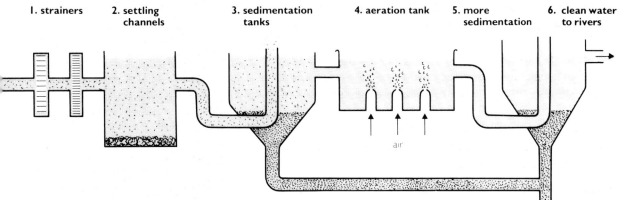

| I. strainers | 2. settling channels | 3. sedimentation tanks | 4. aeration tank | 5. more sedimentation | 6. clean water to rivers |

1. Strainers remove rags, tins and pieces of wood.
2. Grit and sand settle out in channels.
3. The crude sewage flows very slowly through *sedimentation* tanks where about one-third of the organic matter settles at the bottom as sludge.
4. The cleaner water then passes to either *trickling filters* or *aeration tanks* containing sludge agitated with air. Here carefully chosen bacteria live off the sewage breaking down all the remaining polluting matter.
5. More sedimentation tanks collect all the remaining sludge. The clean water is discharged into rivers or the sea.
6. The sludge is collected and warmed in large, sealed tanks. More bacteria *digest* the harmful part of it and produce methane gas and a semi-solid waste. The gas is used at the works to produce electricity: the waste is dumped at sea or dried and used as fertiliser.

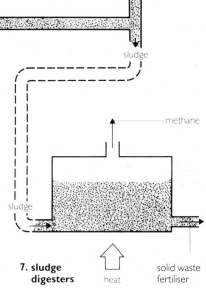

7. sludge digesters

Questions
1. How much water does the body need each day?
2. Give three sources of drinking water. Why does it need treating?
3. Briefly describe the four processes in the treatment of water.
4. What is commonly used to sterilise drinking water?
5. What is sewage? What happens in sedimentation tanks?
6. What happens in aeration tanks?
7. What gas is produced in the sludge digesters? What is it used for?

A sewage plant. The sedimentation tanks are clearly visible. The pits on the left are for sludge storage before final treatment.

Air

We cannot see air but there is a lot of it about – 5000 million million tonnes in all. Apart from keeping us alive – a useful function – the gases in the atmosphere have many other uses.

Gases in the atmosphere

Air is a mixture of gases, mainly nitrogen and oxygen, with small amounts of many others. Water vapour is always present but no percentage is given because the amount varies so much.

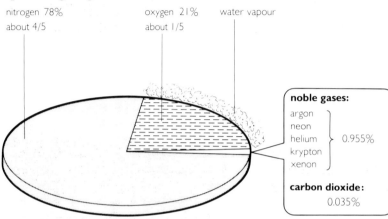

nitrogen 78%
about 4/5

oxygen 21%
about 1/5

water vapour

noble gases:
argon
neon
helium } 0.955%
krypton
xenon

carbon dioxide:
0.035%

Air distillation.

Oxygen

Oxygen is the reactive part of the air. Oxygen enables living things to breathe – it is the gas that combines with materials when they burn. Pure oxygen can be separated from the other gases in air and used to assist breathing in hospitals, aircraft and submarines. Huge quantities are also used in the making, cutting and welding of steel.

Separating oxygen from the air Air is pumped through a filter to remove dust and then carbon dioxide and water are removed. The air is compressed, and then allowed to expand through a narrow jet. This makes it very cold. This cooling process is repeated until the air gets so cold that it changes to liquid.

The extremely cold liquid air contains liquid oxygen, liquid nitrogen and the noble gases in liquid form. This mixture is allowed to warm up slowly and the nitrogen boils away first leaving the others. As the temperature rises the noble gases boil away one by one leaving the oxygen as liquid. The product finally piped away may be high quality gas or liquid or medium purity oxygen gas.

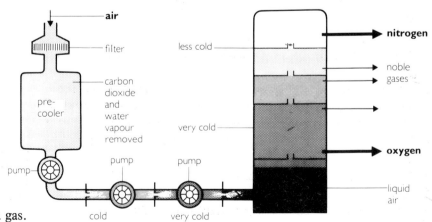

air

filter

carbon dioxide and water vapour removed

pre-cooler

pump

pump

pump

cold

very cold

less cold

very cold

nitrogen

noble gases

oxygen

liquid air

Making oxygen in the laboratory

Some compounds containing a lot of oxygen will release it when they are decomposed (broken down into simpler substances). A solution of hydrogen peroxide gives off oxygen easily and safely. Manganese (IV) oxide is used to make the hydrogen peroxide decompose more quickly. The manganese (IV) oxide is not used up: it acts as a *catalyst*. The reaction is:

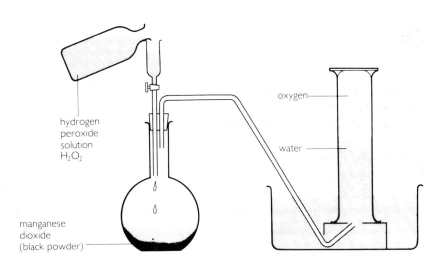

hydrogen peroxide → water + oxygen
$$2H_2O_2 \rightarrow 2H_2O + O_2$$

Nitrogen and the noble gases

Originally the nitrogen produced by liquefying air was wasted because it does not react much. However, this unreactiveness can be a positive advantage. Nitrogen is pumped into oil storage tanks when they are empty, to prevent the possibility of fire.

Liquid nitrogen is very cold. It is used for quickfreezing and transporting food. It can also be used for "shrink fitting". In this process, one very cold metal part is slipped inside another. As it warms up, it gets bigger and fits tightly in place.

Nitrogen can be combined with hydrogen to make the gas *ammonia*, which in turn is used to make nitric acid.

The noble gases   These have no natural reactions at all. Precisely for this reason they find uses in electric light bulbs, in strip lights and in electric arc welding. Helium, being very light, is used for filling balloons.

Producing frozen peas.

Questions

1. List the gases in dry air and give their percentages.
2. What other gas is always present? Why is no percentage given for it?
3. The diagram shows what happens when iron filings are left in a tube with damp air for some days. How far has the water risen in the tube? What proportion is this? What does it show?
4. How is oxygen obtained from the air? Give three uses of pure oxygen.
5. How is oxygen made in the laboratory?
6. Give two uses of liquid nitrogen and two uses of the gas.
7. Give the names of the noble gases and some of their uses.

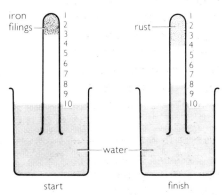

Burning

When a match is lit, the flames give out heat and light energy and the match burns away. Only a thin piece of charcoal is left. Or is anything else formed that we cannot see?

Burning

Burning is the rapid combining of a substance with oxygen, making flames and heat. Air usually provides the oxygen. The main products of burning are *oxides* – the compounds formed when oxygen joins with the elements in the burning substance. These oxides are often invisible gases.

The diagram shows a burning candle in a basin floating on lime water. When covered by a large jar, the candle continues to burn for a short while and then goes out. The lime water rises up the jar – and also turns milky. This shows that only part of the air is used in burning (the one-fifth which is oxygen) – and that carbon dioxide is formed.

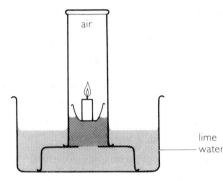

Increase in weight on burning   This can be shown by weighing a bundle of magnesium ribbon in a crucible and then heating it. The lid must be raised from time to time to let air in. Gradually the magnesium changes to the grey ash, magnesium oxide. On reweighing, the magnesium oxide is found to weigh more than the magnesium.

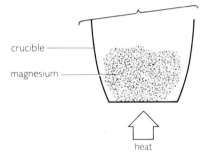

Corrosion

Many metals *corrode* in air: they react slowly with oxygen to form oxides and the surface becomes eaten away. The presence of water greatly assists corrosion. Iron becomes coated with rust in a few days if exposed to both air and water. Dry air or air-free water alone will not cause iron to rust.

Fuels

Fuels are substances which burn easily and release a lot of heat energy. Most fuels are compounds of carbon and hydrogen which give carbon dioxide and water as the main products of combustion. Natural gas is an excellent fuel.

methane + oxygen → carbon dioxide + water + heat
$$CH_4 + 2O_2 \rightarrow CO_2 + 2H_2O$$

Respiration of the body   This is a kind of 'slow burning' or combustion which does not involve flames. The fuel is mainly glucose made from the sugars in food. In the body cells this reacts with oxygen dissolved in the blood to give carbon dioxide and water. Most of the energy released heats the body.

Using one fuel to make another one suitable for use.

Burning of elements in oxygen

Yellow sulphur powder held in a combustion spoon burns with a feeble blue flame when heated. Lowered into a jar of pure oxygen, it burns much more brightly and gives off a gas with a sharp, choking smell. This gas is sulphur dioxide and it will turn a piece of damp litmus paper red.

Many other elements also burn vigorously in oxygen to form *oxides*. Oxides may be put into three groups according to their effect on litmus paper. One group turns litmus red: these are always oxides of non-metals. Another group turns litmus blue: they are oxides of metals. The third group does not affect litmus at all. Water, the oxide of hydrogen, belongs to this group.

Page 39 explains in more detail how the 'litmus red' oxides when dissolved in water make *acids*. 'Litmus blue' oxides, called *bases*, are the 'chemical opposites' of acids.

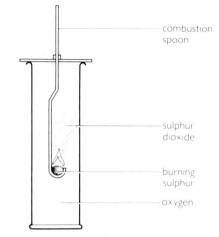

combustion spoon

sulphur dioxide

burning sulphur

oxygen

Element	How it burns	Oxide formed	Effect on litmus
Non-metals			
SULPHUR	blue flame	sulphur dioxide – gas	RED
CARBON	glows red	carbon dioxide – gas	RED
PHOSPHORUS	yellow flame	phosphorus(V) oxide – white smoke	RED
HYDROGEN	blue flame	hydrogen oxide – water	PURPLE
Metals			
MAGNESIUM	bright flash	magnesium oxide – grey ash	BLUE
CALCIUM	red flash	calcium oxide – white solid	BLUE
SODIUM	yellow flash	sodium oxide – yellow solid	BLUE
IRON	glows red	iron oxide – black solid	PURPLE

Testing for oxygen

Things which burn in air will burn much more fiercely in pure oxygen. A smouldering wood splint will burst into flames when put in the gas. This reaction is used as the test for oxygen.

Using pure oxygen gas, this diver can produce a flame hot enough to cut metal underwater.

Questions

1. What is meant by burning?
2. How can it be shown that a candle uses up oxygen when it burns and produces carbon dioxide?
3. Why does magnesium weigh more after burning than before?
4. Name two compounds formed when a fuel burns.
5. Write the equation for the burning of natural gas.
6. Give an example of combustion without flames. Name the fuel.
7. What happens when metals corrode?
8. Name two metal oxides and two non-metal oxides and state how they affect litmus.
9. What is the test for oxygen?

Carbon dioxide

There is only a very small proportion of carbon dioxide gas in the air – 3 parts in 10 000 – and yet all plant and animal life depends on it. Why is it so vital? Will there always be enough of it? How else is it used?

The vital gas

The many different compounds which make up living things are all based on chains of carbon atoms. These are all made from the carbon dioxide in the air. Plants make their food and body cells by absorbing carbon dioxide (through their leaves) and water (through their roots) and changing them into glucose. Oxygen is given off at the same time. Sunlight and the green substance chlorophyll are needed to help the process which is called *photosynthesis*.

$$\text{carbon dioxide} + \text{water} \xrightarrow[\text{sunlight}]{\text{chlorophyll}} \text{glucose} + \text{oxygen}$$
$$6CO_2 + 6H_2O \longrightarrow C_6H_{12}O_6 + 6O_2$$

Some of the glucose is changed into more complex substances, often combined with nitrogen and other elements. Animals eat plants, using the carbon compounds to grow their own bodies.

All living things breathe, turning some of the compounds back into carbon dioxide. All fuels produce carbon dioxide when they burn. The amount of carbon dioxide in the air remains roughly the same because breathing and burning replace the carbon dioxide removed by photosynthesis.

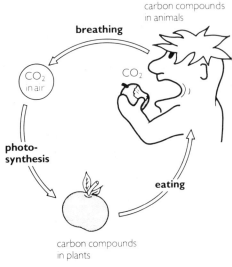

Making carbon dioxide

In the laboratory You can make carbon dioxide by adding moderately dilute hydrochloric acid to chips of marble (a form of calcium carbonate) using the apparatus shown.

$$\text{calcium carbonate} + \text{hydrochloric acid} \rightarrow \text{calcium chloride} + \text{water} + \text{carbon dioxide}$$
$$CaCO_3 + 2HCl \rightarrow CaCl_2 + H_2O + CO_2$$

To obtain dry gas it is passed through a tube containing lumps of calcium chloride. It then pushes the air out of a gas jar.

All carbonates will give off carbon dioxide when treated with acid. Most naturally occurring carbonate rocks will also give off the gas when strongly heated, as explained on page 48.

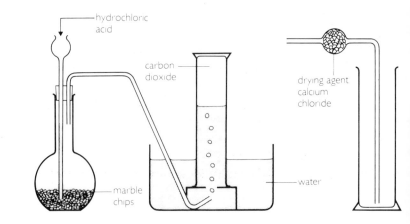

Industrially   Large quantities of carbon dioxide are obtained during the manufacture of hydrogen from natural gas. Carbon dioxide is also made during the production of ethanol (alcohol) when sugar solution is fermented by yeast.

Properties and uses of carbon dioxide
Carbon dioxide has six main properties.

It has no colour or smell.   This makes it difficult to detect.

It does not let things burn in it.   It also changes to a liquid when pressurised. Cylinders of liquid carbon dioxide are used for inflating life rafts, for serving drinks and as fire extinguishers. Being a heavy gas, it forms a blanket over the flames. It is more effective and much less messy and damaging than the liquid produced by the soda-acid type of extinguisher.

Carbon dioxide can solidify when cooled.   The solid is called *dry-ice* and is used for refrigeration. It is especially good during the transport of ice-cream and frozen foods because it keeps them very cold ($-79\,°C$) and changes directly back into gas leaving no mess.

It dissolves in water to give carbonic acid.   This has a pleasant tingling taste and is used to make soda water and fizzy drinks. The same acid, present in rain water, slowly eats away rocks. It can be made to react with a solution of common salt and ammonia to give *bicarbonate of soda* (sodium hydrogencarbonate.) This white powder is used in baking, in health salts and in making washing soda.

Lime water turns milky   when carbon dioxide is bubbled through it. This is used as the test for the gas.

Carbon dioxide changes into carbon monoxide   when heated with carbon. This highly poisonous gas is often present in car exhaust fumes. Carbon monoxide burns in air: it is used as a fuel and in separating iron from the iron ore dug out of the ground.

Carbon dioxide gas may be stored under pressure in cylinders.

Carbon monoxide is produced by car engines even when they are running properly.

Questions
1. What is the proportion of carbon dioxide in the air?
2. How do plants make their food? What is the name of the process?
3. Why is the carbon dioxide in the air not used up?
4. What is marble and what does it give off when hydrochloric acid is added? Write the equation for the reaction.
5. Give two ways of making carbon dioxide industrially.
6. What two properties of carbon dioxide are shown in the diagram?
7. What is 'dry-ice'? Why is it good for keeping things cold?
8. What acid is in fizzy drinks? What is bicarbonate of soda made from?
9. Give the test for carbon dioxide.

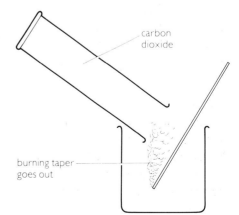
carbon dioxide

burning taper goes out

Further questions

1 **a** Name the two planets which orbit between the Earth and the Sun.
   b Name the planet with rings round it.
   c What orbits between Mars and Jupiter?
   d Complete the sentences:
   Stars produce their own . . . . as a result of . . . . of the gas . . . . A group of stars is known as a . . . . The distance between stars is measured in . . . . The star nearest the Earth is . . . . which is . . . . distant.
   e Scientists always consider the surface of Mars to be more interesting than the Moon. Suggest two reasons why astronauts have been to the Moon but not to Mars. SWEB

2 **a** What name do we give to rock which was once hot and molten?
   b Give one example of this kind of rock.
   c What name do we give to rock which has changed its form as a result of heat or pressure?
   d Give one example of this type of rock. SWEB

3 **a** Which atomic particle has the least mass?
   b Which atomic particle has a positive charge?
   c Name the lightest element. WYLREB

4 The nucleus of an atom of oxygen contains some of the following particles:
   a neutrons, **b** protons, **c** ions,
   d electrons.
   Write down the correct answer. WREB

5 The diagrams below show a simple way of representing atomic particles. Copy and label each diagram correctly: 'uncharged atom', 'negative ion', 'positive ion', 'electron'. SREB

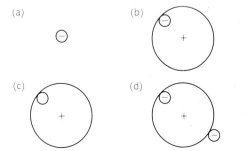

6 **a** Give the chemical symbols for **i** carbon, **ii** calcium, **iii** chlorine, **iv** copper.
   b Sometimes the symbol O may be used, and sometimes the symbol O_2. When used in this way:
   i what does the symbol O stand for?
   ii what does the symbol O_2 stand for? NWREB

7 Give the chemical names of the following.
   a $MgSO_4$, **b** NaCl, **c** $CaCO_3$,
   d Na_2CO_3, **e** NaOH. NWREB

8 The rectangles shown contain symbols arranged to represent molecules of
   a hydrogen and **b** oxygen.

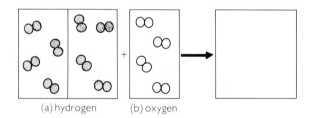

(a) hydrogen   (b) oxygen

Using the same symbols and a box similar to that given, draw in the correct number of molecules representing the new substance formed. EAEB

9 An atom of the element lithium has 3 electrons, 3 protons and 4 neutrons.
   a Make a simple labelled sketch of this atom.
   b What charge is carried by **i** the neutrons, **ii** the protons, **iii** the electrons? EAEB

10 Write out the following table and put a tick in the correct column for each substance.

Substance	Element	Compound	Mixture
sodium chloride			
air			
water			
lead			
soil			
copper sulphate			
diamond			

NWREB

11 a How would you show that ordinary air contains **i** oxygen, **ii** water vapour?
b Name **i** two processes that remove oxygen from the air, and **ii** one process that liberates oxygen into the air.
c The diagram represents a simplified water cycle.

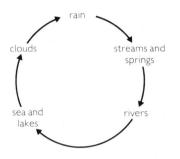

i At which point in the cycle is water most pure? **ii** Why is water from underground sources preferred to rain water for drinking purposes? EAEB

12 a Name three solvents (other than water) which are used in the home or in industry and give one example of the use of each.
b Draw and fully label a diagram of apparatus which could be used to obtain drinking water from sea water.
c What name do we give to the process drawn in **b**?
d Why is very little drinking water obtained in this way?

13 Explain, using simple diagrams where necessary, how the sewerage entering sewerage plants is treated to make it safe to discharge into rivers.

14 Using diagrams, briefly explain how you would carry out these separations:
a Clear water from muddy water.
b Salt from salt water.
c Red and yellow dyes from an orange mixture of these dyes.
d Alcohol from a mixture of alcohol and water.

15 a Briefly describe the industrial extraction of oxygen from the air.
b Give one industrial and one non-industrial use of oxygen.
c What is the approximate percentage of oxygen found in the atmosphere?
d How does the air we breathe out differ from the air we breathe in?
e What is the process called in plants whereby oxygen is given off into the atmosphere? SREB

16 The diagram below represents the approximate composition of dry air by volume. Give the letter which corresponds to **i** oxygen, **ii** nitrogen, **iii** inert (noble) gases, **iv** carbon dioxide. WJEB

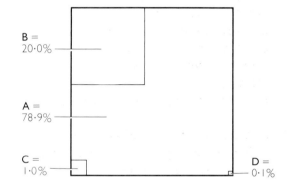

17 Write out the following table and put ticks in the appropriate spaces to show whether the processes listed would increase, decrease, or have no effect on the amount of carbon dioxide in the atmosphere.

Process	Amount of carbon dioxide		
	Increase	Decrease	No effect
Burning carbon			
Photosynthesis			
Transpiration			
Burning magnesium			
Respiration			
Raining			
Burning fuel oil			
Perspiring			

EAEB

18 The two columns below show processes and reactions. For each one of the processes 1–6 there is a matching reaction in A–F.

Process

1. *Displacement*
2. *Fermentation*
3. *Reduction*
4. *Photosynthesis*
5. *Neutralisation*
6. *Oxidation*

Reaction

A Removal of oxygen from a compound

B Acid + alkali → salt + water

C Oxygen combining with another substance

D Magnesium dissolving in dilute sulphuric acid

E Sugar → alcohol

F Carbon dioxide + water → sugar

a For each process state the reaction which best matches it. Use each letter once only.

b i Which process is involved in making larger molecules (which store energy) from smaller molecules? **ii** Where and when does this process take place? <small>LEAG Science</small>

19 Magnesium reacts with dilute hydrochloric acid as follows:

$$Mg(s) + 2HCl(aq) \rightarrow MgCl_2(aq) + H_2(g)$$

a A pupil carried out an experiment to investigate the rate of reaction between a piece of magnesium and $40\,cm^3$ of dilute hydrochloric acid at $20\,°C$. The acid was in excess.
The results are shown on the graph. **i** Copy the graph. **ii** What is the gas produced when

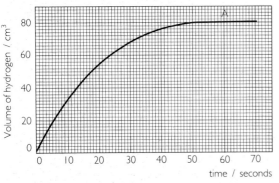

magnesium reacts with hydrochloric acid? **iii** What volume of gas was collected after 30 seconds? **iv** After how many seconds was the reaction just completed?

b The experiment was repeated using an equal mass of magnesium cut into very small pieces and a fresh $40\,cm^3$ of hydrochloric acid at $20\,°C$. The results are shown below.

time (seconds)	0	5	10	15	20	25	30	35	40	45	50
volume of gas (cm^3)	0	34	50	62	70	76	79	80	80	80	80

i Plot these results on to the same paper as graph A. Label the curve graph B. **ii** Which of the two lines refers to the faster reaction? **iii** Explain why the two reactions do not take place at the same rate.

c Which one of the following methods would increase the *final* volume of gas collected? Write the letter of the method you choose and explain your reasoning.

A Use of more magnesium.

B Use of a larger volume of dilute hydrochloric acid.

C Use of a more concentrated solution of hydrochloric acid. <small>LEAG Science (Combined)</small>

20 Spacecraft have to carry all the water that the astronauts need. This chart shows how much water an astronaut uses in one day.

a How much water will need to be put in the spacecraft for each astronaut for each day?

b How much water escapes from an astronaut's body each day?

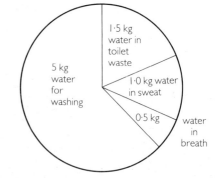

c How much water will an astronaut need to drink each day?

d Different astronauts in fact *don't* all use exactly the same amount of water each day. Does this mean that the figures in the pie-chart are no use? Explain your answer. <small>NEA Modular Science</small>

Chapter 2

Materials from the Earth

Mix the first with the second, and you get the third, which could go on chips. What are they?

The photograph is of 'solid drips'. Hard water?

Acids and bases

The sour taste of unripe fruit, stiffness in tired muscles and the stings of nettles and bees are all due to the presence of natural acids. These acids, and other more powerful ones, can be made artificially. The chemical opposites of acids are called bases.

Acids

Acids can exist as syrupy liquids or solids, but they are most useful as solutions in water. Many oxides of non-metals react with water to give acids, for example:

carbon dioxide + water → carbonic acid
$$CO_2 + H_2O \rightarrow H_2CO_3$$

Acids can be recognised by their reactions with magnesium metal or washing soda (sodium carbonate). Both materials dissolve in acids with a vigorous fizzing.

Acids are corrosive, attacking many materials including flesh! Concentrated acids are very dangerous and should never be handled. Even dilute acids must be treated with care. Keep them well away from the eyes, mouth and any broken skin.

All acids contain hydrogen. Some, or all of this hydrogen is given up when an acid reacts. *Strong acids* release hydrogen very easily whilst *weak acids* release hydrogen with difficulty.

Bases

These are the chemical 'opposites' of acids. Mixed with acids they react strongly, and the acids lose hydrogen. Many oxides of metals are bases. Many bases are insoluble solids; the few which do dissolve in water are called *alkalis*.

Bases are also commonly found in the home and their alkaline solutions are just as corrosive as acids. You should take care not to get alkalis on your skin or clothes. Wash off any splashes with plenty of water.

Frequently-used bases include ammonia solutions in bath and sink cleaners; potassium hydroxide (caustic potash) in oven cleaners; calcium hydroxide (lime) for adding to acid soils; and sodium hydrogencarbonate (bicarbonate of soda) in cooking and medicines.

Vinegar contains acetic acid.

Some more acids in use. Sulphuric, citric, tannic, and tartaric acids are present. Can you sort out which is present in which?

Some common bases. Sodium carbonate is sometimes used when washing clothes; magnesium hydroxide is used to relieve stomach upsets; sodium hydroxide is used to make soap and potassium hydroxide is used in oven cleaners.

Neutralisation

Acids will react with bases to form water and compounds with the general name of *salts*. This reaction is called *neutralisation*.

acid + base → salt + water

Neutralisation is common in everyday life: tooth decay is caused by acids from food and sugar residue; toothpaste is slightly alkaline and counteracts the acid. When the hydrochloric acid in the stomach gets out of balance, sharp pains result. These may be relieved by doses of mild alkali such as bicarbonate of soda or milk of magnesia. Bicarbonate of soda eases acidic bee stings, whereas vinegar (acetic acid) will neutralise the alkaline sting of wasps.

Indicators

Many natural materials have different colours depending on whether they are in acidic or alkaline solutions. These dyes are called *indicators*. You can make one in the laboratory by warming red cabbage leaves with a mixture of water and industrial methylated spirits. Other indicators can be made from beetroots and coloured petals. The most common indicator, litmus, is extracted from lichen. Its colours are:

red – acid;   purple – neutral;   blue – alkaline.

pH paper   This is a sensitive indicator. The paper contains a mixture of dyes which give a range of colours when the paper is dipped into different acids and alkalis. The paper turns red in a strong acid and blue in a strong alkali. Each colour is given a number on the scale 1 to 14. Thus concentrated hydrochloric acid has a pH = 1 and is strongly acidic, whilst acetic acid which has pH = 4 is weakly acidic. Water is neutral and has pH = 7. Sodium hydroxide is a strong alkali and a concentrated solution is about pH = 14. Sodium hydrogencarbonate is very weakly alkaline at pH = 8.

This substance is useful for easing acidic bee stings. What substances shown on the left hand page would be useful for alkaline wasp stings?

Using universal indicator to test for acidity or alkalinity. The colour of the indicator is matched against a printed colour chart.

Questions

1. Name the acid formed when carbon dioxide reacts with water.
2. Name four acids found in foods or drinks.
3. Give two reactions by which acids may be recognised.
4. What element is present in all acids? What is the difference between strong acids and weak acids? How should all acids be treated?
5. What is a base? What is a soluble base called?
6. Give the names and uses of two alkalis found in the home.
7. What are formed when an acid is neutralised by a base?
8. Give two examples of neutralisation.
9. What is an indicator? Give the colours of litmus.
10. What kind of solution has a pH or 4 or 5?

Salts

The salt that we use on our food is sodium chloride. We eat it because it makes food taste better and because our bodies need it. There are hundreds of similar compounds, all called salts. Many are necessary in our diet and others have important industrial uses.

Making salts

All acids contain hydrogen. A salt is formed when hydrogen in an acid is replaced by a metal. Some of the substances which will neutralise acids in this way are shown below.

Acid	Neutraliser	Salt formed	Common name	Other products
sulphuric	magnesium	magnesium sulphate	Epsom salts	hydrogen
sulphuric	sodium hydroxide	sodium sulphate	Glauber's salts	water
hydrochloric	sodium bicarbonate	sodium chloride	common salt	water + carbon dioxide
nitric	ammonia	ammonium nitrate	nitram	—

Making copper sulphate   Copper sulphate can be made in the laboratory by warming dilute sulphuric acid and stirring in small amounts of black copper oxide. The copper oxide dissolves, neutralising the acid and changing it into blue copper sulphate solution. The blue colour darkens until all the acid is used up and no more copper oxide will dissolve. The mixture is then filtered and the clear blue solution is collected in an evaporating basin. Left to stand for several days, the water gradually evaporates off leaving blue crystals of copper sulphate in the dish.

copper oxide + sulphuric acid $\rightarrow$ copper sulphate + water

$$CuO + H_2SO_4 \rightarrow CuSO_4 + H_2O$$

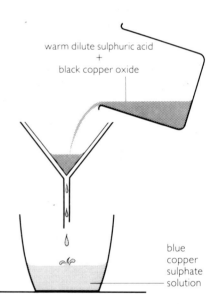

warm dilute sulphuric acid + black copper oxide

blue copper sulphate solution

Natural salts

Rain water dissolves a little carbon dioxide from the air and becomes a very dilute, weak acid. It gradually dissolves rock as soluble salts. Small amounts of these salts are essential to plant growth. Plants absorb them through their roots from the water that is trapped in the soil. Most of the salts are washed away in streams and rivers and collect in the sea. The large deposits of natural salts found in many parts of the world were formed when seas dried up millions of years ago.

Quite often natural salts are found as beautifully coloured and shaped crystals. The photograph shows a strange formation of calcium carbonate crystals. On the opposite page, gypsum crystals are shown, magnified 1000 times.

Properties of salts

Most salts consist of ions (see pages 16–17). Positive (+) ions are provided by the metals, and negative (−) ions come from the acids. For instance, sodium chloride consists of the ions Na^+ and Cl^-. Their ionic nature causes all salts to behave in similar ways.

They are solids with distinct crystal shapes.
Most of them dissolve in water.
When melted or dissolved in water, they conduct electricity.
Because of this they are called electrolytes.

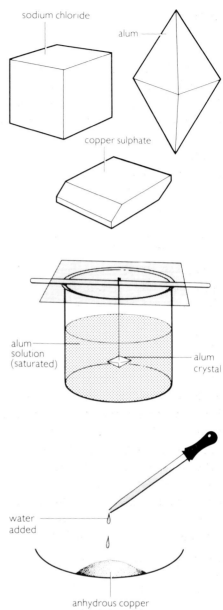

Crystals

Crystals of a salt have a definite shape, which varies with the salt. For example, common salt crystals are simple cubes, and alum crystals are double pyramids. Salts often seem to be powders. If you look at them under a microscope, however, the crystals show up. The picture shows powdered gypsum magnified 1000 times.

Growing crystals Crystals form when a saturated solution of a salt evaporates. The more slowly the water evaporates the larger will be the crystals formed. A crystal of alum can be made larger by hanging it on a thread in a beaker of saturated alum solution. The beaker must be loosely covered to slow down the evaporation and it should be kept at an even temperature. Crystals 2 to 3 cm long will form after several weeks.

Water of crystallisation Many salt crystals contain water: they are called *hydrated* crystals. They give off steam when heated and change to powder. Blue crystals of copper sulphate are hydrated. When heated, they give off their water and turn to a white powder. This is *anhydrous* (dehydrated) copper sulphate. If a few drops of water are added to anhydrous copper sulphate, it goes back to the blue form. This is used as a test for water.

hydrated copper $\rightleftharpoons$ anhydrous copper $+$ water
sulphate (blue)   sulphate (white)   (steam)
$$CuSO_4.5H_2O \rightleftharpoons CuSO_4 + 5H_2O$$

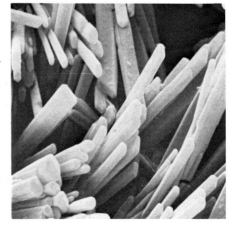

Questions

1. Name three kinds of substances which replace hydrogen in acids.
2. Name the acids from which (a) Epsom salts and (b) washing soda come.
3. Give the chemical names for the two salts in Question 2.
4. Describe how copper sulphate can be made in the laboratory.
5. How are the salts essential to plant growth formed?
6. Why are salts classed as electrolytes?
7. How can large alum crystals be grown? What shape are they?
8. What is dehydrated crystal? What happens when it is heated?
9. Describe a test for water.

Common salt

Sodium chloride or common salt is an essential part of our diet. We need it for our digestive and nervous systems and for our blood. Being so vital, it was one of the first things ever traded. It has even been used as money! Today it is the starting material for many important chemical processes.

How salt is obtained

The huge quantities of salt required by industry come from sea water which contains 2.7% by weight of salt. In Australia and other hot countries, sea water is trapped in large, shallow pools where the sun evaporates the water. The deposits of salt left behind are scooped up by mechanical diggers.

Solid salt is found underground in Cheshire and many other places. The large crystals, coloured brown by impurities, are called *rock salt*. Beds of rock salt were formed when seas of bygone ages dried up. Rock salt is mined like coal and is used in its natural state for de-icing roads and as a fertiliser.

Salt is also obtained by drilling pipes into the salt beds, dissolving the salt in water and pumping the solution (*brine*) to the surface. The salt crystals which remain when the water evaporates are quite pure. Fine crystals for table salt and baking are made by redissolving the salt and evaporating the solution quickly. Other grades of crystals are used for dyeing, leather-making and curing bacon.

Making salt in the laboratory

Sodium chloride crystals can be made by first using a pipette to measure out some sodium hydroxide solution. This is run into a conical flask and a drop of indicator is added. Dilute hydrochloric acid is then added, a little at a time, from a burette. When the solution becomes neutral the amount of acid used is noted. The process is repeated, using the measured amount of acid but leaving out the indicator. Small crystals of salt can be obtained from this solution by boiling off the water.

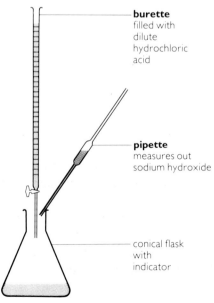

burette
filled with dilute hydrochloric acid

pipette
measures out sodium hydroxide

conical flask with indicator

sodium hydroxide + hydrochloric acid → sodium chloride + water

$NaOH$ + HCl → $NaCl$ + H_2O

Uses of salt

Rock salt   Large quantities of crushed rock salt are used each winter to keep roads clear of snow and ice. A solution of salt in water freezes at a much lower temperature than pure water. Wet roads which have been salted will not freeze up unless the temperature falls to 20 °C below freezing point – much colder than most frosty nights in Britain.

Rock salt is also used as a fertiliser for sugar-beet and wurzels.

Alkali industry   Salt is a cheap and abundant raw material and a great many substances are made from it. Some of these are shown in the diagram:

Sodium is used to make street lamps.

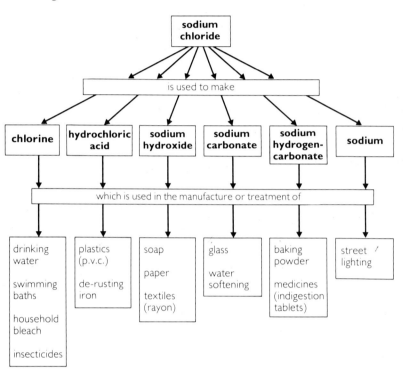

Chlorine is used in swimming pools.

Questions

1. For what parts of the body is salt essential?
2. What percentage of salt is there in sea water?
3. How is salt obtained from sea water?
4. Give two methods of obtaining salt from underground deposits.
5. Give two uses of rock salt and three uses of crystallised salt.
6. Describe how crystals of sodium chloride can be made by neutralizing sodium hydroxide with hydrochloric acid.
7. List five chemicals made directly from salt and give a use of each.
8. List five other substances, found in the home, which come from salt.

Hydrogen

Less than 1% of the Earth's mass is hydrogen. However, hydrogen is a vital part of water and of all living things.

Making hydrogen from acids

All acids contain hydrogen. When metals such as magnesium, zinc, aluminium or iron are put into acids, there is a fizzing as the hydrogen is given off as gas. The metal dissolves to take the place of the hydrogen in solution.

Zinc and dilute sulphuric acid are often used to make hydrogen in the laboratory.

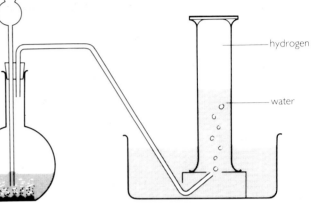

zinc + sulphuric acid → hydrogen + zinc sulphate
$$Zn + H_2SO_4 \rightarrow H_2 + ZnSO_4$$

Properties of hydrogen

Hydrogen has no colour, no smell, and is neutral to litmus. It is much lighter than air – in fact it is the least dense of all gases. If you bubble hydrogen through soap solution, the bubbles rise quickly. They will also burn vigorously if you touch them with a lighted taper.

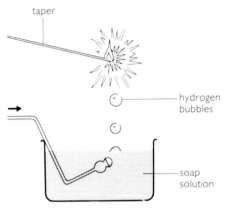

Hydrogen burns very easily and can form dangerously explosive mixtures with pure oxygen. A mixture of hydrogen and air in a tube will explode with a shrill 'pop' when held near a flame. This reaction is used as the test for hydrogen.

The reaction between hydrogen and oxygen produces water. This can be shown by burning a jet of pure, dry hydrogen and playing the flame onto a cool flask. Drops of water may be collected.

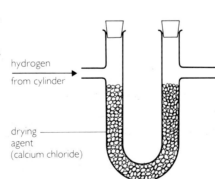

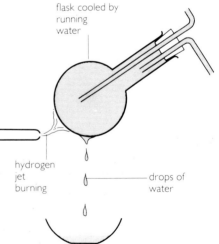

hydrogen + oxygen → water
$$2H_2 + O_2 \rightarrow 2H_2O$$

Water is often formed when substances containing hydrogen react with oxygen or oxygen-containing compounds. The very name 'hydrogen' means 'water producer'.

Hydrogen from water

Water can be made to give up hydrogen in various ways.

Coke Steam passed over red-hot coke produces hydrogen and carbon monoxide gas. This mixture, known as water gas, is used as a fuel and as a source of hydrogen for making ammonia.

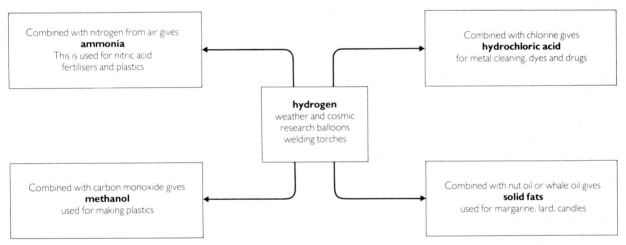

water gas

coke + water → carbon monoxide + hydrogen
$$C + H_2O \rightarrow CO + H_2$$

Very reactive metals Sodium, potassium and calcium are very reactive metals. They combine with water as though it was acid and give off hydrogen gas.

Electricity If an electric current is passed through water containing some acid or salt, the water splits up into oxygen gas and hydrogen gas. This is an industrial method of making hydrogen.

Uses of hydrogen

Hydrogen gas is used for filling balloons, in blow torches for welding and as a fuel for space ships. Large quantities of hydrogen, obtained from natural gas and oil, are converted into a number of other important compounds. Some of these are shown.

Combined with nitrogen from air gives **ammonia** This is used for nitric acid fertilisers and plastics	Combined with chlorine gives **hydrochloric acid** for metal cleaning, dyes and drugs
hydrogen weather and cosmic research balloons welding torches	
Combined with carbon monoxide gives **methanol** used for making plastics	Combined with nut oil or whale oil gives **solid fats** used for margarine, lard, candles

Questions

1. Name three metals which displace hydrogen from dilute acid
2. Write the equation for zinc dissolving in dilute sulphuric acid.
3. Why is hydrogen used for filling weather balloons?
4. Describe the test for hydrogen.
5. How could you show that water contains hydrogen and oxygen?
6. What is water gas and how is it made?
7. Give two ways of making hydrogen in large quantities.
8. Give four compounds manufactured from hydrogen.
9. What is ammonia used for?
10. What does hydrogen react with to form margarine?

A hydrogen filled balloon.

Sulphuric acid

The most commonly used acid both in the laboratory and in industry is sulphuric acid. Four million tonnes of it are made each year in Britain! It is made from natural sulphur or sulphur compounds.

Sulphur

Sulphur is a yellow solid found in many parts of the world where there have been volcanoes. It is one of the few minerals which melt easily and catch fire. This is why it was an ingredient of the first explosive, gunpowder, which is still used in fireworks. Today the main uses of sulphur are the hardening of rubber and the manufacture of sulphuric acid.

Sulphur is extracted from underground deposits by melting it with hot water under pressure and forcing it to the surface with compressed air. It is either allowed to cool and set hard, or is transported in heated containers whilst still liquid.

The Contact process

This is a three stage process which converts liquid sulphur into concentrated sulphuric acid.

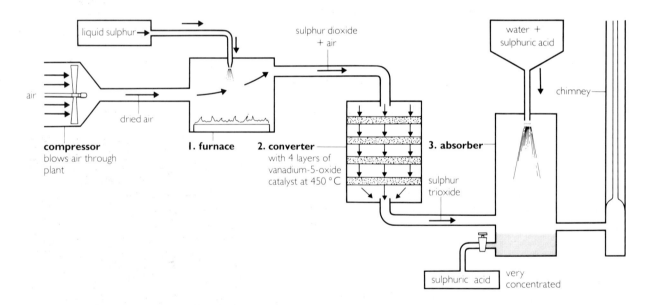

Stage 1: the furnace   Liquid sulphur is burned in a stream of cleaned and dried air, blown through the plant by a huge fan. Sulphur dioxide gas is produced and this passes on to the second stage.

sulphur + oxygen → sulphur dioxide
          (in air)
$$S + O_2 \rightarrow SO_2$$

Stage 2: the converter   Here the sulphur dioxide is combined with more air to form sulphur trioxide. Layers of catalyst pellets, vanadium (V) oxide, heated to 450 °C help the reaction.

sulphur dioxide + oxygen → sulphur trioxide

$$2SO_2 + O_2 \rightarrow 2SO_3$$

Stage 3: the absorber   Here the sulphur trioxide is combined with water to make sulphuric acid.

sulphur trioxide + water → sulphuric acid

$$SO_3 + H_2O \rightarrow H_2SO_4$$

The sulphur trioxide dissolves best in water which already has sulphuric acid in it. The very concentrated acid produced can be diluted as needed.

Uses of sulphuric acid

Sulphuric acid is both a strong acid and a powerful dehydrating agent. Because of these properties it has many uses.

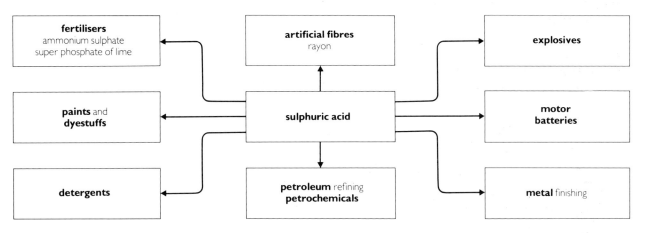

fertilisers ammonium sulphate super phosphate of lime	**artificial fibres** rayon	**explosives**
paints and **dyestuffs**	**sulphuric acid**	**motor batteries**
detergents	**petroleum** refining **petrochemicals**	**metal** finishing

Caution

When mixed with water, concentrated sulphuric acid gives out enough heat to make the mixture boil and spit. This makes it very dangerous. It causes severe burns on skin because of the heat given out as it absorbs water from the flesh. *It should never be handled.*

Questions

1. How much sulphuric acid is made in Britain each year?
2. Why is sulphur an ingredient of gunpowder?
3. How is sulphur obtained from underground deposits?
4. Give two uses of sulphur today.
5. Briefly describe what happens in the three stages of the Contact process for making sulphuric acid.
6. Give the two main properties of concentrated sulphuric acid.
7. List six uses of sulphuric acid.
8. Why is concentrated sulphuric acid very dangerous?

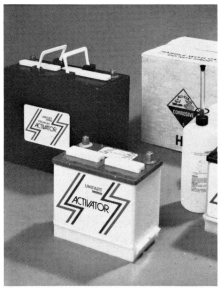

Sulphuric acid is used in car batteries.

Limestone

Limestone is perhaps the commonest rock in the British Isles. In no way a precious stone, it is still an extremely valuable raw material. This section gives some of the reasons why this is so.

Calcium carbonate rock

The term limestone covers all types of calcium carbonate rock. These rocks formed millions of years ago from the shells and bones of creatures which lived in deep, warm seas. You can often see the imprints of shells in limestone rocks: these are called *fossils*. The shell beds were compressed, lifted and folded by later earth movements. They now appear in different forms: carboniferous and oolitic limestones, chalk, marble and calcite.

Quarrying Carboniferous limestone is quarried by removing the soil and blasting the exposed rock with explosives. The large rocks are crushed and graded. Smaller chippings are used for making concrete, iron and glass, and as roadstone and railway ballast. Larger pieces are made into quicklime by heating them strongly in a lime kiln:

A limestone quarry.

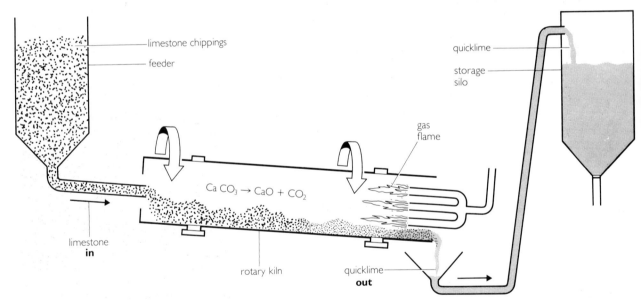

Calcium oxide – quicklime

Heated to 1100 °C, calcium carbonate loses carbon dioxide and changes to the white solid calcium oxide (also called quicklime).

calcium carbonate $\rightarrow$ calcium oxide $+$ carbon dioxide

$$CaCO_3 \rightarrow CaO + CO_2$$

Large quantities of pure quicklime are needed for making steel and other metals. Sodium hydroxide and calcium carbide (from which ethyne (acetylene) gas is made) are also produced from quicklime.

Calcium hydroxide – slaked lime

Calcium oxide reacts vigorously with water, giving off heat and crumbling to a powder called calcium hydroxide or slaked lime.

calcium oxide + water → calcium hydroxide

$$CaO + H_2O → Ca(OH)_2$$

Slaked lime, like quicklime, is strongly alkaline. It is used by farmers to neutralise acid soils, to lighten clay soils and to kill pests. Lime is used to absorb acid gases in the chemical industry and for softening water. Combined with chlorine it forms bleaching powder which is used in the cotton industry.

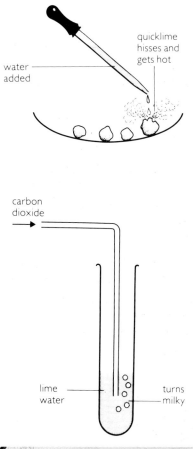

water added

quicklime hisses and gets hot

Lime water

Calcium hydroxide only dissolves slightly in water giving the mild alkali, lime water. This solution absorbs carbon dioxide forming a milky precipitate of pure calcium carbonate.

lime water + carbon dioxide → calcium carbonate + water

$$Ca(OH)_2 + CO_2 → CaCO_3 + H_2O$$

This reaction is the test for carbon dioxide. The 'precipitated chalk' formed is used in paper-making, in toothpaste and in pills.

carbon dioxide

lime water

turns milky

The building industry

The building industry depends entirely upon limestone and limestone products. The stone, particularly the oolitic type, has been used for centuries for buildings of all kinds, from cottages to cathedrals. Roads are made from chippings coated with tar.

Cement   This is a grey powder made by burning limestone, clay and sand in a rotary kiln and grinding the clinker formed with gypsum. Mixed with sand, limestone chippings and water, it sets hard to form concrete, the foremost building material of this century.

Mortar   Brick and stone walls are held together by mortar – a varying mixture of slaked lime, sand, water and cement. Lime mortar sets hard by absorbing carbon dioxide from the air.

The Guggenheim Museum in New York was designed by the famous architect Frank Lloyd Wright. The whole structure is of concrete and steel.

Questions

1. What is the chemical name for limestone? How is it formed?
2. Name three kinds of limestone. What are fossils?
3. Describe how small chippings of limestone are produced.
4. What is quicklime and how is it made? Write the equation.
5. What happens when water is added to quicklime? What is formed? Write the equation for this too.
6. Give three uses each of quicklime and slaked lime.
7. Why does carbon dioxide turn lime water milky?
8. How is cement made? What is mortar?

Hard water

Is ice hard water? No, the term is used to describe water which has calcium or magnesium compounds dissolved in it. This sometimes makes it hard to use.

Soft water

Rain-water is pure because it has nothing dissolved in it except a tiny amount of carbon dioxide. In some parts of the country, tap water is almost pure because the rocks in the area do not dissolve when rain falls on them. This *soft water* lathers easily with soap and leaves no deposit when boiled away. It does not cause radiators or steam irons to block up. Distilled water is the softest of all.

Recognising hard water

Hard water is not pure: it contains calcium salts dissolved out of rocks. These salts give a pleasant taste to the water and are left as a white film if a beaker of hard water is boiled dry. They combine with soap to form a grey scum. Hard water is easily recognised by the large amount of soap it takes to make a lather and by the scum that it leaves. This is what happens:

Distilled water is used to top-up car batteries.

calcium salts + sodium stearate → calcium stearate + sodium salts
(hardness)   (soap)   (scum)

Temporary hardness

There are two kinds of hardness, temporary and permanent. *Temporary hardness* can be reduced by heating. It is caused by calcium hydrogen-carbonate which dissolves out of limestone rocks when rain-water trickles over them.

Rain-water absorbs a little carbon dioxide from the air as it falls. This slightly acid solution attacks the calcium carbonate of limestone, and changes a little of it into soluble calcium hydrogencarbonate. This dissolves easily and the rocks are slowly eaten away leaving pot-holes and caves: the water contains small amounts of limestone in soluble form.

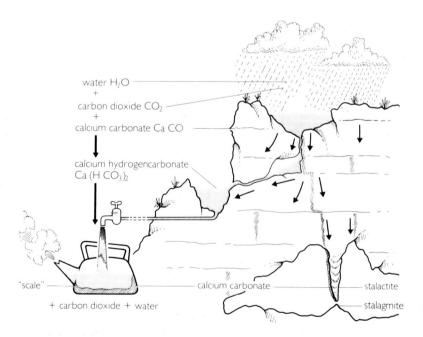

water H_2O
+
carbon dioxide CO_2
+
calcium carbonate $Ca\,CO$

calcium hydrogencarbonate $Ca\,(H\,CO_3)_2$

"scale"
+ carbon dioxide + water

calcium carbonate

stalactite

stalagmite

water + carbon dioxide + limestone ⇌ temporary hardness
H_2O + CO_2 + $CaCO_3$ ⇌ $Ca(HCO_3)_2$

Scaling The reaction between water, carbon dioxide and limestone to produce temporary hardness is *reversible*. When water with temporary hardness in it is heated or allowed to evaporate, the dissolved salt changes back to limestone and is deposited as a solid. This deposit is the 'scale' in kettles. It also builds up as stalactites and stalagmites in limestone caves.

Permanent hardness
Other calcium salts which occur in rocks are calcium sulphate (gypsum) and calcium chloride. Small quantities of these salts dissolve in water to produce *permanent hardness*. Heating this kind of water does not affect the dissolved salts and the hardness remains the same. Magnesium salts also cause permanent hardness.

Removal of hardness
Hard water is often a nuisance because it can use up to five times as much soap for washing as soft water and produces a scum which is difficult to rinse out. Temporary hardness also causes blockages in hot water pipes–wasting heat and making regular de-scaling necessary. For these reasons hard water is often softened by adding substances which remove the calcium salts.

Slaked lime (calcium hydroxide) This cheap material is made from limestone as explained in the last section. It can only remove temporary hardness, converting the soluble calcium hydrogencarbonate back into insoluble calcium carbonate. Slaked lime is added to water storage reservoirs in some limestone areas.

Washing soda (sodium carbonate) This is often used to soften water in the home. It removes all kinds of hardness, changing the calcium salts into insoluble calcium carbonate.

Ion exchange Certain natural substances such as greensand will remove calcium ions from hard water when the water is passed through them. The sand is regenerated from time to time by passing a strong solution of sodium chloride through it. Nowadays synthetic resins from crude oil are used instead of sand.

Using hard water Hard water has some advantages over soft water. The calcium salts help children develop strong teeth and bones. It is also good for brewing beer!

Cross-section through an old water pipe. Use of softened water in heating systems prevents this problem.

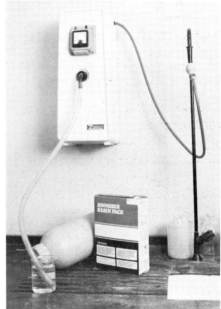
A water softener of the type used in many schools.

Questions
1. What is hard water? Is rain-water hard or soft?
2. Give two ways of recognising hard water.
3. What are the two kinds of hardness? How do they differ?
4. Describe how temporary hardness is formed.
5. What is the 'scale' that forms in kettles? How does it get there?
6. Name two substances which cause permanent hardness.
7. Name two advantages and two disadvantages of hardness in water.
8. Briefly describe two ways of removing hardness.

Ammonia

The smell of a baby's wet nappy is caused by the ammonia gas given off as the urine begins to decay. Far from being an unwanted waste product, ammonia is needed to make fertilisers, explosives and plastics.

Ammonia contains nitrogen

Plants take in nitrogen from the soil and combine it with hydrogen, oxygen and carbon to form cell-building materials called proteins. Animals eat plant proteins, breaking them down into amino-acids. These are re-assembled into new animal proteins. Any over are broken down further into ammonia which is passed out of the body as urea in urine. Buried manure and compost (rotting plants) also release ammonia which is trapped by the soil. In the soil it is converted by bacteria into nitrogen compounds which are absorbed by plants once again. Millions of tonnes of ammonia fertilisers are made and used every year to improve crops worldwide.

Haber process

Ammonia is manufactured by obtaining nitrogen from the air and combining it with hydrogen. The hydrogen comes from water heated with natural gas (methane).

Ammonia is used to make fertilizer.

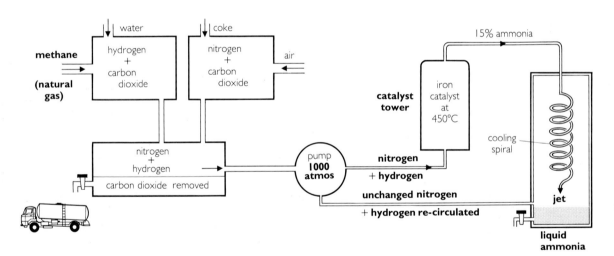

The nitrogen and hydrogen are first pumped up to a very high pressure. They are then passed up a tower packed with a hot iron catalyst. This helps some of the two gases to link together as ammonia.

nitrogen + hydrogen $\xrightleftharpoons{Fe}$ ammonia

$$N_2 + 3H_2 \rightleftharpoons 2NH_3$$

The mixture of ammonia plus unchanged nitrogen and hydrogen are then passed into a cooling chamber. Here the ammonia condenses as a liquid. The unchanged gases are pumped back into the catalyst tower for further conversion to ammonia.

Properties of ammonia

It is an invisible gas, much lighter than air and has a strong smell. It dissolves extremely well in water. A flask full of ammonia gas connected to a trough of water sucks the water up in a fountain.

Ammonia gas turns damp litmus paper blue and damp pH paper a blue/green. It is the only alkaline gas. When ammonia gas mixes with hydrogen chloride gas a white smoke of ammonium chloride particles is formed.

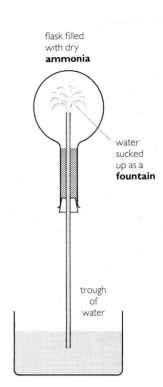

flask filled with dry **ammonia**

water sucked up as a **fountain**

trough of water

Oxidation of ammonia to nitric acid

Much of the ammonia made by the Haber process is converted to nitric acid. This is done by mixing the ammonia with air and passing the gases through heated grids of platinum alloy. The ammonia changes into oxides of nitrogen which are then cooled, mixed with more air, and dissolved in water to give nitric acid.

ammonia + air → nitrogen oxides → nitric acid
$$NH_3 + O_2 \rightarrow NO \text{ and } NO_2 \rightarrow HNO_3$$

Nitric acid is used to make explosives and dyes as well as fertilisers. Nitram fertiliser (ammonium nitrate) is made by neutralising nitric acid with more ammonia.

Uses of ammonia

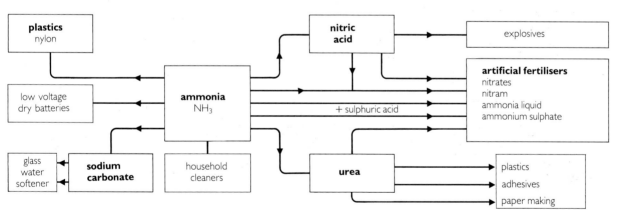

Questions

1. What elements are present in proteins?
2. How do animals produce proteins for their body cells?
3. What happens to unused animal protein?
4. What two conditions are needed to help nitrogen and hydrogen link together as ammonia? Write the equation for the reaction.
5. How does ammonia affect litmus paper?
6. What does the 'fountain experiment' show?
7. What is nitram and how is it made?

Extraction of metals

Metals were the first elements to be used and purified by humans. They are just as important today.

Smelting

Small amounts of some metals can be found lying about in the ground. Gold, silver and copper are the metals most often found uncombined. Meteorites, falling to Earth, have left lumps of iron in some places. Primitive people used these metals for simple tools. Most metals are found combined with other elements as ores. These other elements, usually oxygen or sulphur, can be removed by *smelting* – heating with charcoal. Lead oxide can be smelted to a silvery bead of lead by heating it on a charcoal block. Blowing a bunsen flame on to the ore with a blowpipe helps the carbon to remove the oxygen from the ore. Removal of oxygen from a substance in this way is called *reduction*.

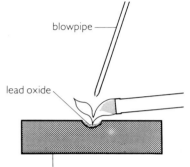

lead oxide + carbon → lead + carbon monoxide
$$PbO + C \rightarrow Pb + CO$$

Tin, lead and copper are all easily extracted from their ores by charcoal smelting. The higher temperature needed to reduce iron ore is produced in the blast furnace.

The blast furnace

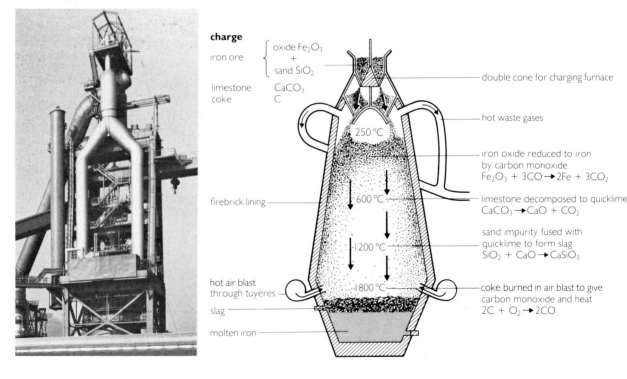

charge

iron ore { oxide Fe_2O_3 + sand SiO_2 }

limestone $CaCO_3$
coke C

double cone for charging furnace

hot waste gases

250 °C

iron oxide reduced to iron by carbon monoxide
$Fe_2O_3 + 3CO \rightarrow 2Fe + 3CO_2$

firebrick lining

600 °C

limestone decomposed to quicklime
$CaCO_3 \rightarrow CaO + CO_2$

sand impurity fused with quicklime to form slag
$SiO_2 + CaO \rightarrow CaSiO_3$

1 200 °C

hot air blast through tuyères

1 800 °C

coke burned in air blast to give carbon monoxide and heat
$2C + O_2 \rightarrow 2CO$

slag

molten iron

Activity league

Metals are alike in many ways. They are shiny and strong: they conduct heat and electricity. However, they differ in their ability to form compounds. Metals at the top of the activity table react so well that their compounds are extremely stable. These ores can only be split up to give the metal by using powerful currents of electricity.

Metal	How found – ore		How reactive	How obtained
potassium sodium	sylvine rock salt	KCl NaCl	Extremely reactive	Metals can only be separated from their ores by using very large currents of electricity
calcium magnesium aluminium	limestone magnesite bauxite	$CaCO_3$ $MgCO_3$ Al_2O_3	Very reactive	
zinc iron	calamine haematite	$ZnCO_3$ Fe_2O_3	Moderately reactive	Metals formed by heating strongly with carbon
tin lead copper	cassiterite galena chalcopyrite	SnO_2 PbS $CuFeS_2$	Feebly reactive	These metals are easily extracted by heating with carbon
silver	argentite	Ag_2S	Unreactive	Metals produced by heat alone.
gold	pure metal			

Electrolysis of bauxite

Aluminium is one of the metals which is extracted by electrolysis. The cell illustrated is one of about 200 used in an aluminium production plant.

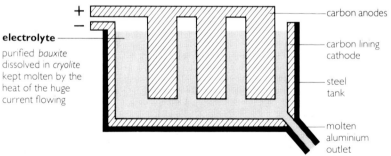

Aluminium ore being reduced to the metal by electrolysis.

Questions

1. Name three metals which occur uncombined in some places.
2. What is meant by (a) an ore, (b) smelting?
3. Name three ores which are easily smelted.
4. What does haematite consist of? Explain how it is reduced to iron in the Blast Furnace and how slag is formed.
5. Give three ways in which metals are similar.
6. Name a very reactive metal and one of its ores.
7. Describe how aluminium is extracted from bauxite.

Uses of metals

Gold is used as an international currency – other pure metals are extremely valuable too. But mixtures of metals are often more useful than pure metals for making things.

Pure metals

Pure gold is used as an international currency. Being unreactive and easy to shape, it has always been prized for ornaments and jewellery. Other pure metals used today are: lead for pipes and roofing; zinc for batteries; aluminium for saucepans, foil wrapping and milk bottle tops; copper for electrical wiring, pipes and hot water cylinders.

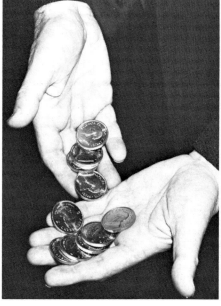

Two pure metals in use – lead and gold.

Pure metals are generally too soft for engineering purposes. The hardness, strength and resistance to corrosion of a metal are increased by mixing it with another metal to form an *alloy*.

Copper alloys

Copper is one ingredient in three important alloys:

Coin metal   Pure metal coins quickly wear out because the metal is too soft. 'Silver' coins used to be alloys of silver with copper but are now nickel and copper alloys.

Bronze   This alloy of copper and tin was first made about 5000 years ago by smelting mixed ores of copper and tin. It is harder and stronger than pure copper, and is easily melted and cast. It was the first metal to be widely used for weapons and tools. It is still used for statues, coins and cutlery and also for gun metal (with zinc), springs (with phosphorus), pumps and propellers (with aluminium) and for wheel and engine bearings (with lead).

Brass   is made by dissolving solid zinc in molten copper and solidifying the alloy in sand moulds. Like all alloys its composition can vary – a typical brass will be 70% copper and 30% zinc. Brass is strong, does not corrode and can be die cast (pressed into shape when cold). Shell cases, electrical fittings, car radiators, and some nuts, bolts and screws are made of brass.

Bronze is used to make church bells.

Iron alloys

Iron is a very important substance, but is often used as an alloy rather than pure.

Pig iron is produced in a blast furnace. It contains about 4% carbon and smaller amounts of other elements. The high carbon content makes this alloy very hard – it is useful for castings such as cylinder blocks and cookers. Unfortunately it cannot be welded and snaps under strain.

Steel has a much lower carbon content (between 0.1% and 1.5%), and is much stronger than pig iron. It is made by burning away most of the carbon impurity from molten pig iron using a jet of pure oxygen. *Mild steel* (up to 0.5% carbon) is by far the most important metal. It can be rolled out into sheets, made into girders, stamped into shapes and welded together. *Hard steel* (0.5 to 1.5% carbon) is used for tools because cutting edges remain sharp.

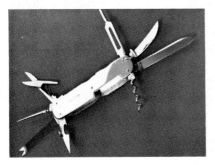

Two alloys in use – different types of steel for penknives and bridge cables.

Special steels contain other metals in addition to the carbon. Chromium in *stainless steel* adds strength, prevents rusting and gives an attractive finish. Ballbearings and drills are made from stainless steel besides the familiar cutlery, sinks and teapots. *Manganese steel* is extremely hard and tough and is used for rock-crushing, drilling and for railway points.

Lead alloys

Lead forms alloys with low melting points. Mixed with tin, it forms *solder* for electrical connections and joints in pipes. Lead/antimony/tin alloys are used as *type-metal* for printing and as *white metal* for wheel bearings.

Questions

1. State uses of four different pure metals.
2. What are alloys? Why are they preferred to pure metals?
3. What is a 'silver' coin made of?
4. State the metals present in bronze and five uses of bronze.
5. How is brass made? What is die casting?
6. How is pig iron converted into steel?
7. Give three uses each of mild steel and hard steel.
8. What elements does stainless steel contain?
9. Name three lead alloys and their uses.

In this machine, a bar of typemetal is gradually being melted and poured into moulds which will form it into the letters required.

Corrosion

The bright shine of a freshly cleaned metal soon goes dull (tarnishes) because the metal reacts with air and forms a layer on the surface. This layer can be used to protect some metals. If the metal continues to react with air then corrosion sets in and the metal is slowly eaten away. Corrosion of iron is a serious problem.

Air attacks metals

All metals react with the oxygen in air to some extent. Calcium and other metals high in the activity table (see page 55) join with oxygen very rapidly to form oxides. They corrode away completely.

calcium + oxygen → calcium oxide
$$2Ca + O_2 \rightarrow 2CaO$$

Sodium and potassium are so reactive that they have to be stored in oil. They tarnish and corrode very quickly if exposed to air. Metals such as aluminium and zinc are in the middle of the activity table and they join with oxygen more slowly. A layer of oxide does form after some time and it sticks firmly to the surface. The rest of the metal is protected by this layer and so does not corrode any more. Aluminium is protected from corrosion by *anodising* – using an electric current to deposit a thickened layer of aluminium oxide onto the surface.

Metals at the bottom of the table oxidise very slowly. But even gold changes in time from light yellow to the browner shade of 'old gold'.

Silver has to be cleaned regularly because it tarnishes by combining with sulphur compounds in the air. Copper, used for roofing, also joins slowly with sulphur compounds producing a green coating or *patina*.

For safety, sodium is stored under oil.

Water attacks metals

A small piece of potassium reacts violently when put into water. It fizzes around and bursts into lilac coloured flames. Metals below potassium in the table react less vigorously. With calcium it is safe to trap and test the gas given off. The gas is hydrogen.

water + calcium → hydrogen + calcium oxide
$$H_2O + Ca \rightarrow H_2 + CaO$$

The calcium oxide changes to calcium hydroxide and makes the remaining solution strongly alkaline.

Less reactive metals will only combine with water in the form of steam. Iron, heated in steam, forms a protective coating of black magnetic iron oxide (Fe_3O_4).

Metals low in the table do not react with water much. Lead has been used for water pipes since Roman times.

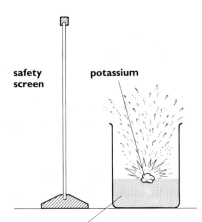

safety screen

potassium

water – becomes an alkali (potassium hydroxide)

Rusting of iron

By far the most widely used metal is iron – usually in the form of steel. Its only fault is that it corrodes so badly and forms rust. Rust is hydrated iron oxide (iron oxide plus water) and it forms when air and water together attack iron.

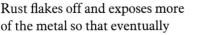

Rust flakes off and exposes more of the metal so that eventually the iron will rust right through. Rusting occurs because iron has areas of slightly different composition. These form an electric cell (page 214) when covered by water which has carbon dioxide and oxygen dissolved in it. A tiny current flows and some of the iron dissolves. Rusting is quicker if the water is more acidic or if it contains salt.

Protecting iron

Keeping it dry   A layer of oil or grease stops water from reaching tools or the moving parts of machines. Tar is also a good protector of iron which is not seen or handled. A bag of silica gel drying agent can be put inside the cases of instruments like balances or typewriters to protect the iron.

Painting   To prevent rust, the steel is treated with a rust inhibitor like phosphoric acid, then coated with a zinc based paint.

Plating   Plating with other metals stops corrosion. *Galvanised iron* is made by dipping acid-cleaned iron into a bath of melted zinc. The layer of zinc gives good protection even if scratched. When a scratch occurs zinc, iron and water form an electric cell. Since the zinc is more active it dissolves away first leaving the iron rust free. Tin cans are made of steel coated with tin. Tin is not affected by food acids. However, if the tin layer is damaged, the iron rusts more than ever because it is higher in the activity table than tin.

Cathodic protection   Many underground gas and water pipes are made of steel. They cannot be repainted so they are joined by wires to blocks of magnesium alloy. The magnesium, steel pipes and water in the soil form an electric cell. The more active magnesium corrodes instead of the steel and can be replaced when used up. This is also used to protect ships' hulls and pipes in oil refineries.

The primer paint used on this bridge contains zinc compounds.

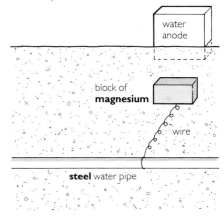

Questions

1. Name a metal that reacts quickly with air. What is formed?
2. Explain what is meant by 'anodising'.
3. What happens when potassium is put into water?
4. What is rust? Why is it such a problem?
5. Give details of four different methods of protecting iron.

Fossil fuels

Anything that burns is a fuel. Coal, oil (petroleum) and natural gas are called fossil fuels because they have been buried deep underground for millions of years. They are the remains of plants and animals and consist mainly of compounds of carbon and hydrogen (hydrocarbons).

Burning of fossil fuels

When fuels burn, their hydrocarbons combine with oxygen in the air forming carbon dioxide and water. Flames are produced and heat is given out. This heat is used to warm houses, for industrial furnaces, to drive motors and to make electricity.

hydrocarbons + oxygen $\rightarrow$ carbon dioxide + water + heat

A certain amount of the poisonous gas, carbon monoxide, is produced when the air supply to a burning fuel is limited. It is the deadly gas present in car engine exhaust fumes.

Coal

Most of the coal in Britain came from the swampy forests that grew 250 million years ago. Trees and giant ferns grew quickly and died forming thick layers of rotting trunks and leaves called *peat*. The swamps sank slowly and the peat layers were covered with mud and sand. Seas then flooded the land, gradually covering the peat and mud with the shells of dead sea creatures.

A fossil fern in a lump of coal.

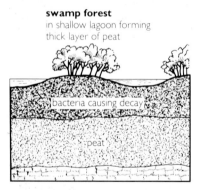

swamp forest
in shallow lagoon forming thick layer of peat

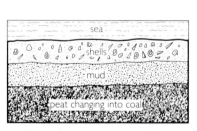

area submerged
deep under the sea

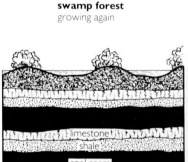

swamp forest
growing again

After thousands of years, the seas retreated and swamp forests again grew in the tropical conditions. The cycle was repeated many times, over millions of years, and the layers were pushed down deeper and deeper. Because of the heat and pressure, the peat lost much of the hydrogen and oxygen in the original woody tissue and hardened into coal seams.

Types of coal There are three main types of coal which differ in age and in the alteration of the original peat.

Lignite (70% carbon) is brown, soft and woody.
Bituminous (85% carbon) is older, black and breaks easily.
Anthracite (93% carbon) is oldest, shiny black and very hard.

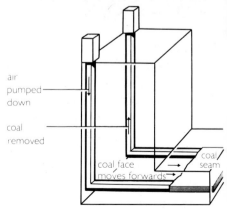

Oil (petroleum)

Crude oil is a sticky, smelly liquid, dark brown or greenish in colour. It comes from the decayed remains of sea creatures and plants which were trapped millions of years ago in the mud and sand of the sea bed. The decay process used up most of the oxygen present in the rotting bodies leaving hydrogen and carbon compounds. As the mud was buried and hardened by later deposits, the hydrocarbons were squeezed out. They moved upwards through soft, porous rock and reached the surface in some places. In other places, such as the North Sea, the hydrocarbons lie trapped under layers of solid rock shaped like domes (anticlines). In these oilfields, as they are called, the hydrocarbons are contained in soft rock like water in a sponge.

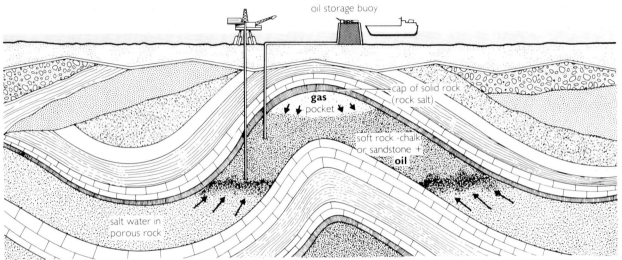

oil storage buoy

gas pocket

cap of solid rock (rock salt)

soft rock - chalk or sandstone + **oil**

salt water in porous rock

Natural gas

Methane and a small proportion of other simple hydrocarbons separate from the oil as natural gas. There is usually a pocket of natural gas under the dome in an oilfield. The pressure of this gas forces the oil to the surface if shafts are drilled into the oilfield. In some fields only the gas has been trapped. This natural gas is piped ashore and supplied direct to consumers through a network of underground pipes. Natural gas is now used as a source of hydrogen which is used in turn to make ammonia and nitric acid.

Questions

1. Name the fossil fuels and their main constituents.
2. What do fuels form when they burn?
3. When was Britain's coal formed?
4. Explain the formation of coal seams.
5. Name three types of coal. How do they differ?
6. What is crude oil and how was it formed?
7. What are the usual features of an oilfield?
8. Name the main hydrocarbon of natural gas and one substance made from it.

Laying a pipe for North Sea Gas supply.

Coal and oil products

Look around you at any time and you will see many things made from coal and oil – clothes, furniture, house fittings, car parts and all sorts of containers.

Distillation of coal

When coal is strongly heated with no air present, it does not burn but breaks down (distils) into four main products:

coal gas – once used for heating;
ammonia – made into fertilisers;
tar – drugs, dyes and plastics;
coke – still used for furnaces and for extracting metals.

Although natural gas and oil have replaced coal gas and tar at the present time, this process may be of major importance again when the world's oil supplies run out.

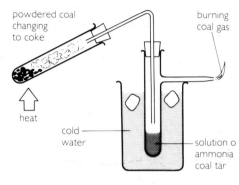

Distillation of oil

The compounds of carbon and hydrogen that make up crude oil are separated by distillation into *fractions* (groups of hydrocarbons with molecules of similar size).

In the laboratory fume cupboard   five fractions can be obtained using the simple apparatus shown. On gentle heating a watery liquid distils over. As the heating is increased the fractions become darker and more oily. Finally a yellow/brown grease collects.

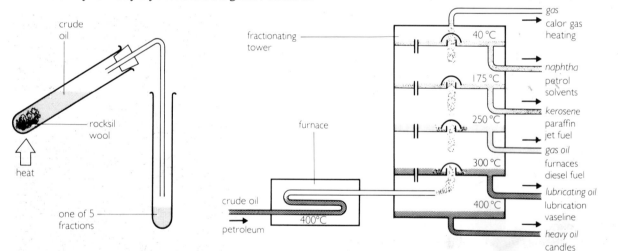

In an oil refinery   crude oil is heated to 400 °C to turn all the components into gases. These pass into a fractionating tower where they turn back into liquids at different levels. The heavy oils condense at the bottom of the tower and the light fractions condense at the top. Each of the fractions is further refined by having impurities removed and by being distilled again to give a range of products.

Cracking of oil

Crude oil contains only a small proportion of the more useful lighter fractions such as petrol. Some of the heavier oil fractions are turned into petrol and jet fuel by the process of *cracking*. The large oil molecules are broken down into smaller petrol molecules by heating with steam or with a catalyst. As an example:

heavy oil $\xrightarrow{\text{heat}}$ petrol type compound + gas (propene)

$$C_{11}H_{24} \longrightarrow \qquad C_8H_{18} \qquad + \qquad C_3H_6$$

The propene gas produced during cracking is made into plastics. The gases ethene and butene are also produced by cracking. They too are used as raw materials in the plastics industry.

Plastics

Plastics are tough, synthetic materials which are easily moulded by heat and pressure. Celluloid was one of the first plastics, invented over 100 years ago, and is still used for table tennis balls. Many plastics used to be made from coal tar, but now the cracking of oil provides most of the starting materials. All plastics are made by joining together small molecules called *monomers* to make very big ones – made of hundreds of carbon atoms linked into chains. The giant molecules are called *polymers* and the linking process is *polymerisation*.

Polythene is made by causing lots of molecules of ethene gas to join together. This can be done by heating them under pressure with aluminium and titanium compounds acting as catalysts.

Other plastics based on materials produced from ethene include: PVC (drain pipes, clothing); PVA (glue, emulsion paint); polystyrene (heat insulation).

Nylon and polyester are polymers made from more complex products of oil refining. Their long chains are arranged side by side, giving them properties like those of cotton fibres.

The plastic 'Alkathene' in use for road markers.

Questions
1. Name the four products of the distillation of coal.
2. Describe how crude oil is separated into fractions industrially.
3. List six final products of an oil refinery.
4. Why is coal gas no longer used?
5. What is cracking? Name the gases formed as by-products.
6. What is polymerisation? Explain how polythene is made.
7. Name two other plastics based on ethene and give a use for each.
8. Why do the properties of nylon make it suitable for clothing?

Detergents

Water, by itself, is not a good washing agent. It does not get grease off a plate or stains off a handkerchief without a great deal of rubbing. We rely on detergents to help.

Detergent action

Detergents are substances which help water to squeeze between the fibres of cloth and which also react with dirt. Detergent molecules have water-attracting heads and long, grease-attracting tails. The first problem the detergent has to overcome is *surface tension* which holds the water into small droplets, as explained in more detail on page 74. The water-attracting heads break this down, helping the water to penetrate the cloth.

Second, the grease-attracting tails of the detergent molecules attach themselves to the grease. They form layers which pull the grease away from the cloth, holding it as a suspension in the water. Because of this, the grease can be rinsed away, and harsh rubbing of the cloth becomes unnecessary.

Solid particles of dirt are also removed by detergents. The process is similar to grease removal but double layers of detergent molecules form around the dirt particles. These double layers pull the dirt away from the cloth.

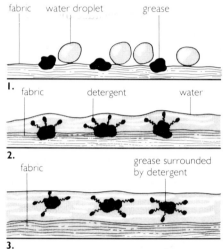

Soap

Soap, the first detergent, was originally made from animal fats combined with alkali obtained from wood ashes. Soap can be made

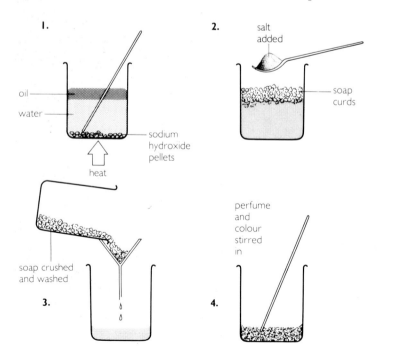

This machine presses soap into moulds in the industrial manufacture of soap.

in the laboratory by stirring cooking oil or fat with sodium hydroxide solution and boiling it for an hour.

The large scale manufacture of soap is based on the same principle. Peanuts are one source of the oils boiled with sodium hydroxide to make the sodium salts which are soap. Glycerine is formed at the same time, and this is separated and used to make medicines and explosives.

Soapless detergents

Soap is expensive because the vegetable oils from which it is made are very costly. Soap does not deal very well with heavy grease and it combines with the calcium salts in hard water forming a messy scum. Scientists experimented for many years to produce detergents which did not suffer from the disadvantages of soap. Since 1950, a variety of soapless detergents have been developed, mostly based on substances obtained from crude oil.

Long chain hydrocarbons from oil provide the tails of the detergent molecules. The heads come either from sulphuric acid or from other oil products. Sodium salts are often added: sodium phosphate to loosen the dirt and sodium perborate as bleach. *Biological* washing powders have enzymes added instead of bleach to digest stains.

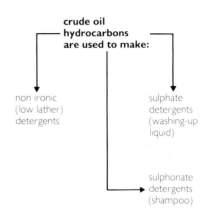

crude oil
hydrocarbons
are used to make:

non ironic
(low lather)
detergents

sulphate
detergents
(washing-up
liquid)

sulphonate
detergents
(shampoo)

Soft detergents

Soapless detergents are such good lathering agents that rivers receiving used water containing detergents can become covered with thick layers of foam which cause pollution. Foam can also cause trouble in sewage works. 'Soft' or *biodegradable* detergents have been developed to avoid this problem. Their molecules break down when attacked by the usual bacteria that cause decay.

The problem of "foaming rivers" has largely been solved since the introduction of biodegradable detergents.

Questions

1. State two ways in which detergents help water to wash things.
2. Describe the structure of a detergent molecule.
3. Describe how detergent molecules remove grease.
4. Describe how soap can be made in the laboratory.
5. Give three disadvantages of soap as a detergent.
6. What are biological washing powders?
7. What are biodegradable detergents?

65

Further questions

1 Put these stages in the preparation of copper sulphate crystals into the right order:
   a Unused copper oxide powder filtered out.
   b Liquid heated to remove some water.
   c The mixture is warmed.
   d Black copper oxide powder is added to dilute sulphuric acid.
   e Slow cooling of the remaining liquid gives crystals of copper sulphate.
   f Resulting clear solution is blue in colour.
   g Excess copper oxide added to use up all the acid.

2 Name the substances formed when the following chemicals react together:
   a Hydrochloric acid and sodium carbonate;
   b Sulphuric acid and sodium hydroxide.   EAEB

3 Write out the following table putting a tick in the correct column.

All acids:	True	False
a turn litmus red;		
b burn clothing;		
c contain hydrogen;		
d have a sour taste;		
e react with bases to form salts;		
f contain oxygen.		

4 The following crystalline substances are commonly found in the home: *washing soda*, *common salt* and *Epsom salts*. Give their chemical names and also the name of a crystalline substance which changes colour on heating.   EAEB

5 Which of the following substances is used
   a as a fertiliser   b to soften water
   c to make soap   d on food?
   sodium chloride, sodium carbonate, ammonium sulphate, sodium hydroxide, acetic acid.   WMEB

6 The following diagram, which is incomplete, shows the relationship between some compounds of calcium. Copy the diagram into your book and then answer the questions below.

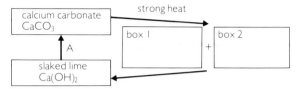

a In boxes 1 and 2 fill in the names and formulae of the compounds which are formed.
b One of the compounds you have named in a can be converted to slaked lime. How can this be done?
c What is the correct chemical name for slaked lime?
d What do we call the solution which is formed when a small quantity of slaked lime is dissolved in water?
e Slaked lime in solution can be converted to calcium carbonate as shown by arrow A. How can this be done?
f Give one common name for calcium carbonate as it occurs in its natural state.
g Give one important commercial use for calcium carbonate.   ALSEB

7 a Write a word equation to show the action of heat on calcium carbonate.
b Describe in detail the action of rain water on rocks containing calcium carbonate. In your answer include any reactions which occur and name any compounds formed.
c List the materials used in the manufacture of cement.
d Describe the manufacture of cement.   EAEB

8 a Draw and fully label a large diagram of a blast furnace.
   b i Name one ore of aluminium.
     ii Describe the extraction of aluminium from the ore.   iii Give 2 uses of aluminium in buildings.   EAEB

9 Why is aluminium extracted by electrolysis and not in a blast furnace? WMEB

10 Lead can be obtained by the reduction of lead oxide on a charcoal block.
    a Explain the term reduction.
    b What is the reducing agent in this reaction?
    c What is being oxidised in the reaction? WMEB

11 **a** What is an alloy?
    b What is the difference in composition between steel and iron?
    c Select the alloys from the following list: lead, iron, solder, tin, brass, zinc, copper, steel and silver.
    d Give one use for lead.
    e Describe one method by which you could extract lead from lead oxide in the laboratory. Give the equation for the reaction in words or symbols. WYLREB

12 Which of the following metals is most likely to be found in its natural state? Tin, lead, iron, gold, calcium, sodium. WYLREB

13 State 2 physical properties and 2 chemical properties of a named metal. EAEB

14 **a** How would you do a very simple test to find out whether or not a material is a metal?
    b State one physical property of gold and one chemical property of zinc.

15 What are *fossil fuels*? Where does the energy in fossil fuels originally come from?

16 **a** What is the chief element found in coal?
    b If this element were burned in plenty of oxygen, what would be formed?
    c What is the test for the substance produced in **b**?
    d How would a solution of the substance formed in **b** react with litmus? YREB

17 Describe how natural gas was formed and say in what conditions it may be found in quantity. EMREB

18 Describe a laboratory experiment in which you could collect samples of at least 3 of the main products of the destructive distillation of coal. Draw a diagram of the apparatus and say in what ways each of the products obtained in your experiment are important to us. EMREB

19 Name 3 fraction (products) of the distillation of petroleum (crude oil). ALSEB

20 What are the properties and uses of 2 of the following polymers?
    a nylon
    b polythene
    c P.V.C. WREB

21 Describe **a** how you would prepare a sample of soap and **b** how you would test its pH value. EMREB

22 Detergent dissolves in water to give a neutral solution. Explain how a household detergent acts on dirt particles on cotton fibres to remove the dirt during the washing process. (Labelled diagrams could be used). SREB

23 **a** Zinc and iron are metals. Both react with dilute hydrochloric acid. **i** Name 2 properties which you would expect metals such as zinc and iron to have. **ii** Explain how you would find out which of these 2 metals reacted most quickly with dilute hydrochloric acid.
    b copper, iron and aluminium are economically important metals. Give an example in each case of an every-day use.
    c The figures in the table show the percentage corrosion (reaction) of some steel bars placed in solutions of various acidities, measured as units of pH.

% corrosion of steel bars	65	60	55	50	20	15	10
pH of solution	1	2	3	4	5	6	7

i Draw a graph of % corrosion against pH using the figures in the table. **ii** A solution of pH4 causes 50% corrosion of the steel bars. Explain what you would do to change the pH to give a solution which was less corrosive.

LEAG – Science 1988

24 a All metallic elements are good conductors of electricity. Most non-metals are insulators.
i State one large scale use of a named metallic element based on its electrical conductivity.
ii State 2 other physical properties which are typical of metallic elements.
b Complete the following word equation for the reaction of a metallic element with an acid.

Acid + metal → .... + ...

c Two test tubes X and Y were set up as shown and after a week it was noticed that X showed considerable rusting while in Y rusting was slight.
i State one reason for the difference noted. **ii** What would be seen if the stopper was removed from tube Y and it was left for another week? Explain your answer.

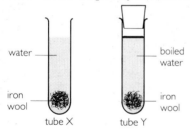

d Objects made of steel may be protected from rusting by dipping in molten zinc.
i What is this process called? **ii** These objects remain protected even if the coating is scratched. Why is this?
e State 2 different methods for protecting parts of a bicycle from rusting.
f Suggest one reason why cans used for food storage are coated with tin.

SEG – Science 1988

25 The manufacture of sulphuric acid can be represented by the following flowchart.

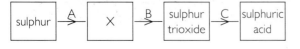

a Stage A involves the burning of sulphur.
i Give the name of the main substance formed (box X). **ii** Write a chemical equation for this reaction.
b Vanadium (V) oxide is a catalyst for the reaction.
i What effect does vanadium (V) oxide have on the reaction? **ii** State one other general property of a catalyst. **iii** Sulphur trioxide is not dissolved directly in water at stage C. Describe how stage C is carried out.
c Sulphur dioxide is produced in most power stations in Britain as part of the waste gases. Explain what effect this gas may have on:
i living organisms **ii** buildings.
d Suggest a method of removing sulphur dioxide from the waste gases. LEAG – Science (Combined)

26 a During refining, crude oil is split up into fractions having different physical properties and uses. The diagram shows an industrial plant suitable for separating a sample of crude oil into fractions.

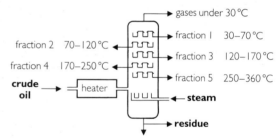

i What name is given to the separation process? **ii** Which of the fractions catches alight most easily? **iii** Fraction 5 has a higher boiling point than fraction 1. Describe 2 other differences you might expect in the physical properties of these fractions. **iv** State which fraction would be best for each of the following uses: camping gas
lubricating oil
petrol
road surfacing.

b Discuss 2 problems caused by the accidental release of oil at sea (in the form of oil slicks) and outline a method of dealing with oil slicks.

LEAG – Science (Combined)

Chapter 3

Molecules and waves

What's the connection between sunbathing, listening to the radio and swimming in the sea?

The photograph shows a blockage of wave energy. What is it?

Hot molecules

Scientists have devised a theory to explain why solids, liquids and gases behave in the way they do. No-one can prove that the theory is true but all the evidence seems to support it. It's called the kinetic theory of matter.

Solids, liquids and gases

According to the *kinetic theory of matter*, solids, liquids and gases are made up of tiny particles called *molecules* which strongly attract each other when they are close together. These molecules are constantly on the move:

Solids In a solid, strong forces of attraction pull the molecules close together. The molecules move by vibrating from side to side.

Liquids In a liquid, the forces of attraction still pull the molecules into the smallest possible space but the molecules have enough freedom to be able to move about as they vibrate – the liquid can flow.

Gases In a gas, the molecules are well spaced out and virtually free of any attractions. They move about at a very high speed and quickly fill any space available to them. As they move they collide with each other and with the walls of any container they happen to be in.

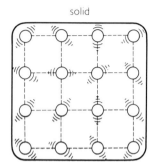

solid

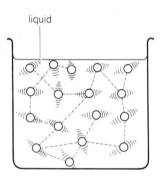

liquid

Heating solids, liquids and gases

As a substance gets hotter its molecules move faster. If the molecules of a solid are heated they begin to overcome the attractions which hold them together and the solid may become a liquid.

If the molecules of a liquid are moving fast enough to completely break free of attractions, the liquid becomes a gas.

Water is one common substance you will recognise in all three forms – as a solid called ice, as a liquid, and as a gas known as steam.

If ice is heated sufficiently it melts to form liquid water.

If liquid water is heated sufficiently it boils to form steam.

A liquid tends to change into a gas long before it is hot enough to boil but the change happens most rapidly at the boiling point.

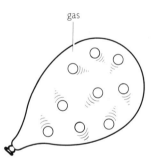

gas

Changing state

Water can be made to change state quite easily, from solid to liquid to gas and back again. Not many other familiar substances do this. The concrete in buildings does not melt on a hot day, and the air does not liquefy on a cold one!

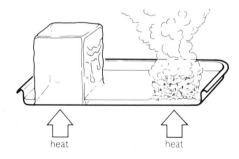

heat   heat

Temperature: the Celsius scale

The degree of hotness of a solid, liquid or gas is measured using a *scale of temperature*. The most commonly used scale of temperature is the *Celsius* or Centigrade scale; the numbers on this scale are called '*degrees Celsius*' and written °C. On this scale:

pure ice melts at 0 °C;
pure water boils at 100 °C.

The chart below gives some typical temperature values.

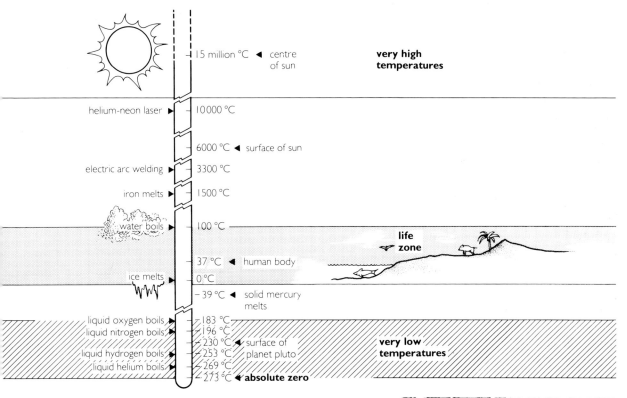

Absolute zero According to the kinetic theory, molecules move more slowly as they get colder. At a certain temperature, −273 °C, all movement stops. This temperature is called *absolute zero* and it is the coldest temperature possible. (It is now known that molecules have *least* movement at absolute zero, rather than no movement at all.)

Questions

1. Describe the movements of the molecules in (a) a solid (b) a liquid (c) a gas. Are molecular attractions least in a solid, a liquid or a gas?
2. What happens to molecules as they become hotter and hotter?
3. What names are given to water in its solid, and gas forms?
4. At what temperature does ice melt on the Celsius scale?
5. At what temperature does water boil on the Celsius scale?
6. At what temperature does liquid oxygen boil?
7. What is absolute zero? What is special about this temperature?

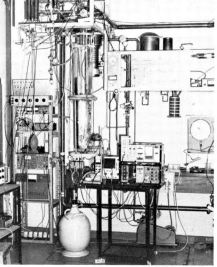

'Cold molecules'. This equipment is used to investigate low temperatures.

Wandering molecules

The kinetic theory can be used to explain several strange effects that it would be difficult to account for otherwise. Here are a few examples.

Brownian motion

Smoke is made up of billions of tiny particles of ash. Viewed through a microscope, these particles can be seen to wander about in a jerky zig-zag way. The effect is called *Brownian motion* and it happens because the smoke particles are light enough to be moved when individual molecules in the air crash into them.

The diagram shows the apparatus used to show the Brownian motion of smoke particles.

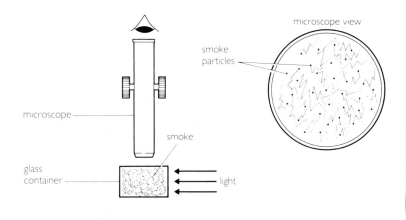

Some liquid dye has been carefully added to still water in the beaker.

Diffusion

If a small amount of coloured dye is dropped into a beaker of water, the colour soon spreads throughout the water. The molecules in the dye spread by a process called *diffusion* – they wander at random through the water as they are bumped and jostled by the molecules around them.

Diffusion will cause a small amount of leaking gas to spread throughout a room. It also causes smells to spread – smells are molecules which come from whatever it is that happens to be smelling.

After twelve hours the dye has diffused upwards through the water.

Osmosis

The experiment on the right shows a process called *osmosis* in action.

The tube contains concentrated sugar solution (a mixture of sugar molecules and water molecules) but the beaker contains water only. The sugar solution and the water are separated by a thin sheet of Visking tubing – called a *membrane* – which is stretched across the mouth of the funnel.

At first, the sugar solution and the water are at the same level – but the level of the liquid begins to rise as water passes through the membrane and into the sugar solution. The sugar solution becomes more dilute as the height of the liquid increases.

The process by which water passes through the membrane is called osmosis. The pressure which builds up above the membrane is called *osmotic pressure*.

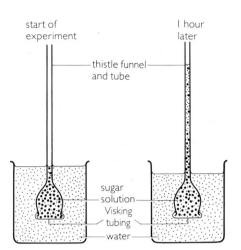

How the kinetic theory explains osmosis   The Visking tubing has tiny holes in it which are so small that the sugar molecules are too large to wander through. But the water molecules are able to. They go through the membrane – though more go into the sugar solution than out of it because more are in contact with one side of the membrane than the other (see diagram).

Semi-permeable membranes   Materials such as Visking tubing which allow some substances to pass through but not others are described as *semi-permeable*. Water will move by osmosis through any semi-permeable membrane that separates a weak solution from a strong one.

The cells in the roots of plants absorb water from the soil by osmosis.

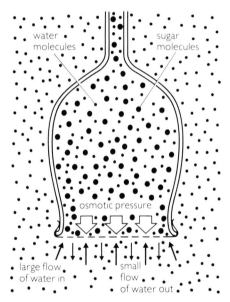

Osmosis in cells   Semi-permeable membranes form the outside of the tiny liquid-filled cells from which animals and plants are made. They are useful in a variety of ways, as suggested below:

Osmotic pressure inside plant cells keeps them firm rather as balloons are kept firm by the air pressure inside them. The firmness of the cells keeps plants upright – without water, they wilt.

Dried fruit swells when soaked in water because cells in the fruit absorb water by osmosis.

Questions

1. Why do the smoke particles in the experiment on the opposite page appear to dance about? Name these movements.
2. What is diffusion? Give three examples of diffusion.
3. What happens to the sugar solution in the experiment described above? Why?
4. A thin sheet of Visking tubing is semi-permeable. What does this mean?
5. Give two examples of osmosis in action in plants.

The effect of reduction in osmotic pressure.

Surface tension and its effects

A drop of water behaves as if its surface was covered with a stretchy skin. The effect is caused by attractions between molecules – it makes life easier for mosquitoes and pond skaters but harder for house builders!

Surface tension

The fact that there is tension in the surface of a liquid can be demonstrated using soapy water and the wire frame shown in the diagram. The frame is first dipped in soapy water so that a thin film of liquid stretches across it.

If the film to the right of the thread is popped, the film remaining pulls the thread firmly to the left.

Surface tension pulls drips from a tap into round droplets and it makes mercury collect in round blobs. It enables you to float a needle on the surface of water without it sinking and it allows insects to stand on the surface of ponds.

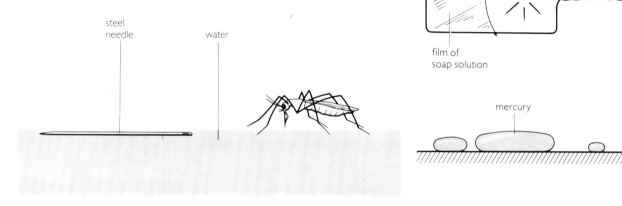

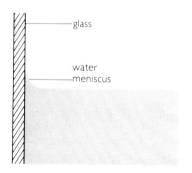

Cohesion

Cohesion   A liquid does not really have a stretchy skin. The tension in its surface is caused by the attractions which pull all the molecules together. Attraction between molecules of the same type is known as *cohesion*. These attractions will pull small amounts of a liquid into round droplets. They will also resist attempts by needles and the feet of insects to push molecules apart by breaking through the surface of water.

Adhesion   Molecules of a liquid may also be attracted to molecules of other materials – an attraction of this type is known as *adhesion*.

If you empty water out of a beaker, drops of water stay clinging to the glass. This is because the water molecules are more strongly attracted to the glass than they are to each other – the adhesion of water to glass is stronger than the cohesion of water.

The strong attraction of water molecules to glass makes the surface (the *meniscus*) of water curve upwards where it meets glass.

Capillary attraction

Glass attracts water molecules strongly enough to pull a thin column of water up through a narrow tube (a *capillary* tube) made of glass. The effect is called *capillary attraction*. The narrower the tube, the further the water rises, because the weight of water needing support is less.

Water, alcohol and many other liquids can be drawn up through narrow gaps by capillary attraction.

Materials like blotting paper and cotton wool which 'soak up' liquids contain thousands of tiny air spaces through which liquids can move by capillary attraction. Soil and porous rocks also contain air spaces and they absorb water in the same way.

In the walls and floors of houses, capillary attraction would cause rising damp through the bricks and concrete if steps were not taken to prevent it – there are air spaces in concrete and in brickwork.

In new houses a waterproof sheet of polythene is laid in the concrete base to stop the upward movement of water from the ground. For the same reason, a waterproof layer is set into the outer brick walls just above ground level. This waterproof layer is called a *damp course* and it is made from a mixture of felt and bitumen.

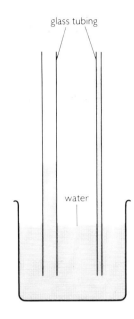

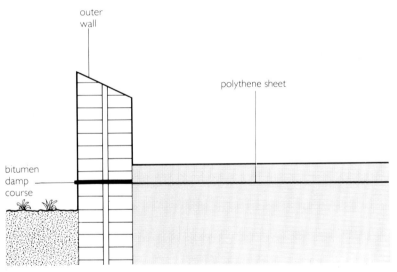

A pond skater demonstrates the effect of surface tension.

Questions

1. How could you show that there is tension in the surface of a liquid?
2. Give three examples of the effects of surface tension.
3. What is cohesion? How is it different from adhesion?
4. Why do drops of water cling to the sides of a glass beaker?
5. What happens if a narrow glass tube is dipped in water? What would be the effect of using an even narrower tube?
6. Give two practical examples of capillary attraction.
7. What is a damp course? What is its purpose?

Expanding solids

Solids expand when they are heated – by so little that you would not normally notice it. Yet an expanding solid can produce enough force to crack concrete or buckle steel. And it all happens because vibrating molecules need room to move.

How solids expand

When a solid is heated its molecules vibrate more vigorously than before. As the vibrations become larger the molecules are pushed further apart and the solid expands.

Solids do not expand very much when heated but the forces produced by their expansion can be extremely high. It is possible to stop a solid expanding but only by clamping it very firmly to something rigid.

Problems with expansion

Concrete and steel beams are often used in construction work but they can cause considerable damage if they do not have enough room to expand when the weather gets hot.

Gaps are left at the ends of bridges to give room for expansion. One end of a bridge is often supported on rollers so that movement is possible as the bridge expands.

Railway lines are laid down in very long lengths. Any expansion in the rails is resisted by fastening them to heavy concrete sleepers embedded in chippings – this prevents buckling in hot weather. Some expansion does take place at the ends of each length of rail, but the lengths are linked by overlapping joints which allow movement.

100 m of steel

When 100 m of steel is heated 10 °C........ ....it expands 11 mm

Expansion in different materials

A 100 metre length of steel railway line heated through 10 °C expands only 11 millimetres.
The expansion would be lower still if the line were shorter or its temperature rise were less.

The table on the opposite page shows how much different materials would expand compared with the length of steel.

Expansion of 100 m length of material heated through 10 °C			
Brass	19 mm	Invar (metal)	1 mm
Iron	12 mm	Concrete	11 mm
Steel	11 mm	Glass	9 mm
Platinum alloy	9 mm	Pyrex glass	3 mm

The information in the table above is useful when selecting materials to do particular jobs:

Steel rods can be placed in concrete beams to strengthen them because the steel and the concrete will expand equally in hot weather. If the expansion of each was different, the steel rods might crack the concrete.

Platinum alloy wires sealed in glass are used to carry the electricity into a light bulb. When the bulb heats up the expansion of the platinum alloy will be the same as that of the glass, so there is no risk of the glass cracking.

Pyrex glass expands less than ordinary glass and so is much more use in the kitchen. If boiling water is poured into a glass dish, the inside of the dish begins to heat up and expand a few moments before the outside. As the inside of the dish pushes against the outside the glass may crack. A pyrex dish is less likely to crack because its glass does not expand so much.

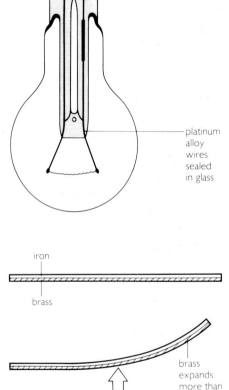

platinum alloy wires sealed in glass

The bimetal strip

Thin strips of brass and iron can be bonded together to form a *bimetal strip*.

If a bimetal strip is heated, the brass expands about 1½ times more than the iron. This makes the bimetal strip bend with the brass on the outside – the longer side – of the curve.

The bimetal strip has many uses, some of which are described in the next section.

iron

brass

brass expands more than iron

heat

Questions
1. Why do solids expand when heated?
2. How is expansion allowed for in a bridge? Where else might expansion be a problem?
3. In the table at the top of the page:- Which material expands least? Which material expands most?
4. Which material expands the same amount as concrete? Why is this useful?
5. What advantage does pyrex glass have over ordinary glass?
6. What happens to a bimetal strip when you heat it? Why?

Using expansion

Whilst expansion is often a nuisance, it can be very useful too.

Joining metals using expansion

Expansion can be used to produce a tight fit between two pieces of metal. In the diagram below, a steel tyre is being fitted to the rim of a train wheel.

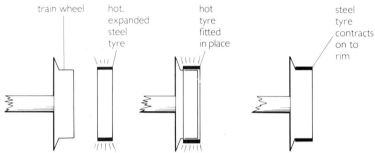

To begin with, the inside of the steel tyre is slightly smaller than the rim of the wheel. The tyre is then heated until it expands enough to fit loosely over the rim. When the steel tyre cools, it contracts (gets smaller) and tightly grips the rim of the wheel.

The bimetal thermometer

A bimetal strip can be used to make a simple thermometer. The bimetal strip is in the form of a long spiral and the centre of this spiral is attached to a pointer.

When the temperature rises, the bimetal strip coils itself into an even tighter spiral and the pointer is moved across the scale.

The bimetal thermometer is less accurate than many other thermometers but it is tough and easy to read.

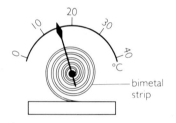

The thermostat

Thermostats are devices which control temperature. Ovens, immersion heaters and refrigerators are all fitted with thermostats – so too are some electric room heaters.

The thermostat in the diagram is connected to an electric room heater. As the room warms up, the bimetal strip bends and the two electrical contacts are moved apart. This switches off the heater. As the room cools down, the bimetal strip cools and straightens. The heater is switched on again as the contacts touch. In this way, the thermostat switches the heater on and off to keep the room at a more or less steady temperature.

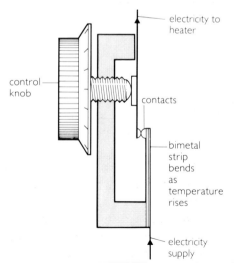

The temperature is selected by turning the control knob. If the control knob is screwed inwards, the bimetal strip has to bend further before the contacts separate – the room needs to be warmer before the heater is switched off.

Flashing indicators

The indicator bulbs on many cars flash on and off because of movements made by a tiny bimetal strip. The diagrams below show one simple system in which a bimetal strip makes a light bulb flash on and off.

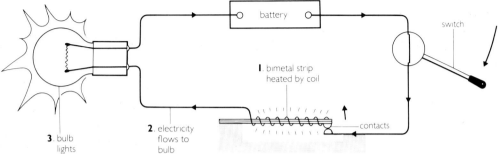

1. bimetal strip heated by coil

2. electricity flows to bulb

3. bulb lights

switch

battery

contacts

When you turn the switch on, electricity flows from the battery to the bulb and the bulb lights up. On its way, this electricity also flows through a small heating coil wound round a bimetal strip. Heated by the coil, the bimetal strip starts to bend. As it bends, it moves one of the contacts away from the other.

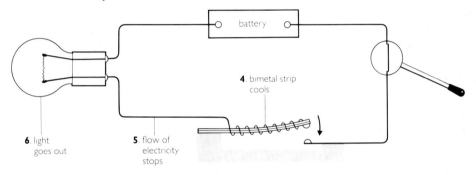

battery

4. bimetal strip cools

6. light goes out

5. flow of electricity stops

With the contacts apart, no electricity can reach the bulb or the heating coil – the light goes out and the bimetal strip cools and straightens.

The contacts touch again as the bimetal strip becomes straight – electricity once more flows through bulb and coil, and the whole process repeats itself.

In this way, the light bulb continues to flash on and off until you turn the indicator switch off.

Questions

1. Why is a steel tyre heated before being fitted to a train wheel?
2. What is a thermostat used for? What pieces of equipment might a thermostat be fitted to?
   How does the thermostat in the diagram switch off electric current? If the control knob is screwed outwards, what effect will this have?
3. When you first turn on the switch in the diagrams above, the bulb lights up. Why does the light then go out? Why does the light come on again after this?

Expanding liquids

Just like solids, liquids expand as they get hotter. In fact, they expand very much more than solids, and their high rate of expansion makes them very useful in thermometers.

Liquid-in-glass thermometers

Mercury thermometers   A typical mercury thermometer is shown in the diagram on the right.

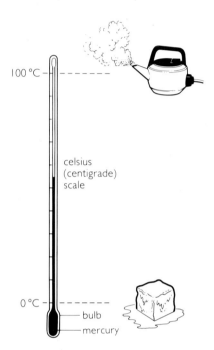

When the temperature rises, the expansion of the mercury in the bulb pushes the mercury 'thread' further up through the narrow glass tube. The tube is the same width from top to bottom – it is made narrow so that a small expansion of the mercury produces a large upward movement of the thread.

To mark an accurate temperature scale on the thermometer, two points on the scale are first fixed:

The thermometer must read 0 °C when placed in pure melting ice.

The thermometer must read 100 °C when placed in the steam above pure boiling water (the water must be boiled under standard atmospheric conditions).

The clinical thermometer   Doctors and nurses use a mercury thermometer with two special features:

The thermometer only covers a narrow range of temperatures close to the average body temperature of 37 °C but it can measure temperatures in this range very accurately.

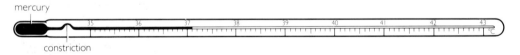

The tube has a narrow bend (a constriction) in it to stop the mercury thread moving back down the tube after the thermometer is removed from the patient's mouth. The temperature can be read any time after this, but the mercury then has to be shaken back down the tube.

Alcohol thermometers   Alcohol is sometimes used in thermometers rather than mercury. It expands about six times as much as mercury for the same temperature rise and is less poisonous should the thermometer break. An alcohol thermometer can also be used in arctic temperatures that might freeze mercury. Mercury freezes at −40 °C, alcohol at −115 °C.

Disadvantages of alcohol are that it needs to be coloured to be seen easily and that it tends to cling to the side of the tube when the thread is falling.

The maximum and minimum thermometer A maximum and minimum thermometer is shown in the diagram on the right. It is an alcohol thermometer with two small steel pointers to record the highest and lowest temperatures reached over a period of time.

When the temperature rises, the expanding alcohol pushes the mercury thread further along the glass tube, and the mercury thread carries the pointer to a higher position on the maximum scale. The pointer sticks in the tube at the highest point reached. Similarly, when the temperature falls, the mercury thread moves back along the tube and the other pointer is carried further along the minimum scale. Both pointers can be pulled back again with a small magnet.

The expansion of water

Water behaves in a most unusual way when heated from 0 °C. As its temperature is raised from 0 °C to 4 °C, water actually *contracts*. From 4 °C upwards, water expands like other liquids.

Water therefore takes up least space at 4 °C – it has its greatest *density* at this temperature. It will sink down through warmer or colder water around it.

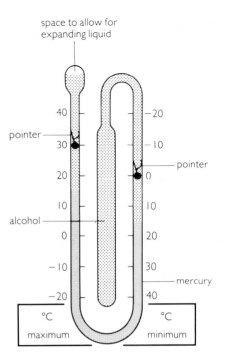

Water may still be at 4 °C at the bottom of a lake even though conditions on the surface are freezing. Fish and other forms of water life are able to survive a severe winter by staying in this deeper water.

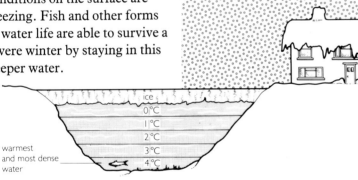

Questions

1. Which expand most when heated, liquids or solids?
2. Why is the tube in a liquid-filled thermometer narrow?
3. How is a clinical thermometer different from an ordinary mercury thermometer?
4. What are the advantages of using alcohol rather than mercury in a thermometer? What are the disadvantages?
5. What were the maximum and minimum temperatures recorded by the pointers on the thermometer shown on this page?
6. What happens to water when it is heated from 0 °C to 4 °C?
7. Why might fish survive winter at the bottom of a lake?

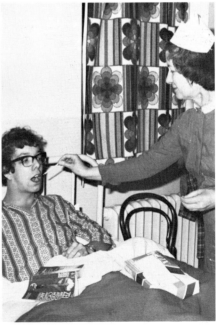

A clinical thermometer in use.

81

Conduction

It is very difficult to keep heat in one place – the vibrations of molecules always try to spread out as much as possible. Home insulation isn't keeping the cold out, but the vibrations in!

How materials conduct heat

When a material is heated, its molecules begin to move more quickly. Each molecule bumps against those around it and eventually all the molecules speed up. In this way, heat is conducted to all parts of the material.

In metals, some heat is conducted as above but most is conducted in another way.

Some of the electrons in the metal atoms are free to drift about between the atoms. When a metal is heated these *free electrons* speed up and carry their heat to all parts of the metal.

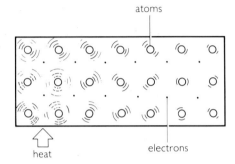

Good conductors of heat

Most metals conduct heat well because they contain free electrons – these same electrons also make metals good conductors of electricity.

Of all the metals, silver and copper are the best conductors.

Good conductors often feel cold to touch. A metal bar feels cold when you pick it up because it rapidly conducts heat away from your hand.

good conductors
most metals especially:
silver
copper
aluminium

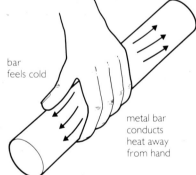

bar feels cold

metal bar conducts heat away from hand

Uses of good conductors

Good conductors have many uses: saucepans are often made of aluminium because it readily conducts heat from a gas ring or a hotplate. The bases of saucepans are sometimes made of copper for the same reason.

If wire gauze is placed over a bunsen flame, the gauze conducts heat away from the flame. This cools the hot burning gas so that no flame appears above the gauze. A glass beaker can be safely heated on the gauze because this protects it from the concentrated heat of the flame.

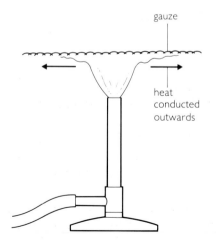

gauze

heat conducted outwards

Poor conductors of heat

Non-metals tend to be poor conductors of heat (and electricity) because none of their electrons are free to move about.

Materials such as cork, plastic and asbestos are poor conductors of heat, so are most liquids. Gases are even worse than liquids at conducting heat. Many materials act as poor conductors because they contain trapped air.

Materials which are poor conductors of heat are known as *insulators*.

Uses of insulators   Insulating materials are useful in the kitchen, to stop heat going where it isn't wanted. For example:

Table mats are often made of cork to protect table tops from hot dishes. Saucepan handles are often made from plastic so that as little heat as possible is conducted to your hand.

There are many examples of trapped air acting as an insulator.

Furry animals keep warm because their fur traps air. Even snow with air trapped in it can act as an insulating blanket. Woollen jumpers are warm because they trap air, so are the fillings used in sleeping bags. Fibreglass and polyurethane foam are good insulators because of the air trapped inside them.

poor conductors (insulators)
water, most liquids
cork
PVC and other plastics
glass
wood
asbestos
air and other gases
materials which
trap air:
wool
fur
fibreglass
plastic foam

Insulating the house   The diagram shows a house in which a number of insulating materials have been used to cut down losses of heat.

Although glass is a poor conductor of heat, a single sheet of window-glass is thin enough for heat losses through it to be noticeable.

A *double-glazed* window consists of two sheets of glass with an insulating layer of air trapped between them.

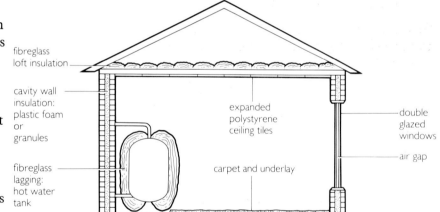

fibreglass loft insulation

cavity wall insulation: plastic foam or granules

fibreglass lagging: hot water tank

expanded polystyrene ceiling tiles

carpet and underlay

double glazed windows

air gap

Questions

1. How is heat conducted through a metal?
2. Name three good conductors of heat.
3. Which materials would you recommend for a table-mat, the base of a saucepan, the handle of a kettle?
4. Which are usually the poorest conductors of heat, liquids or gases?
5. What makes fur and fibreglass good insulators?
6. What is a double-glazed window? Why is it useful?

Natural insulation.

Convection

If liquids are such poor conductors of heat why does a kettleful of water warm through so quickly?
If gases are such poor conductors of heat why is the air above a bunsen flame so hot?
This section looks at the way in which heat can be carried by a rising stream of liquid or gas – the process called convection.

Convection in a liquid

The circulating stream of water in the diagram is called a *convection current* – the potassium permanganate crystals colour the water so that the current can be seen easily.

The current circulates as warm water rises above the heat of the flame and cooler water sinks to take its place. The water expands when heated and is pushed upwards by the cooler, denser water around it.

The domestic hot water system   Most household water systems make use of convection to circulate hot water through the pipes. The diagram shows the layout of a simple system – in many houses the storage tank is in an upstairs cupboard and the header tank is in the loft.

Water heated in the boiler rises by convection to the top of the storage tank. Cooler water in the bottom of the tank sinks back down to the boiler where it is heated. In this way, a supply of hot water collects in the storage tank from the top downwards.

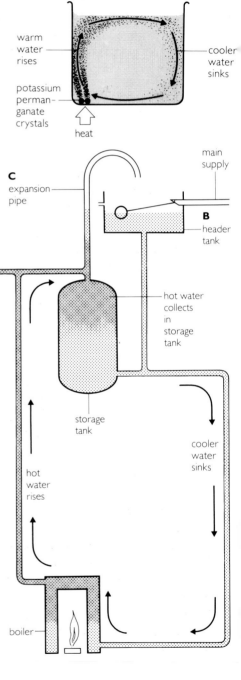

The *hot taps* (A) are supplied with water from the top of the storage tank.

The *header tank* (B) provides the pressure needed to push hot water out of the taps. It also replaces water in the system as it is used. When the water level in the header tank drops, the plastic float also drops – this opens a valve to let in more water from the main water supply. As the tank fills, the float rises and the supply of water is shut off.

The *expansion pipe* (C) serves as an overflow should steam or air bubbles build up in the system.

Convection in air

When convection takes place in air:
hot air rises and *cooler air sinks*.

Warming a room by convection   Many room heaters, including 'radiators', are designed to circulate warm air by convection.

The diagram shows a convector heater in use. The heater is placed close to the floor and warm air rises above it. A convection current is set up as cooler air sinks into the bottom of the heater. In time, all the air in the room is warmed as it circulates through the heater.

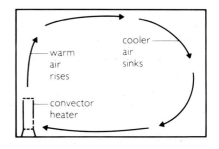

Convection in a refrigerator   Cold air is circulated round the food in a refrigerator by convection. The freezer compartment is placed near the top of the refrigerator so that cold air sinks away from it. Warmer air rises to the top of the refrigerator where it is cooled by the freezer compartment.

Land and sea breezes   Convection causes the onshore and offshore winds which sometimes blow on the coast during the summer.

In hot sunshine, the land heats up more quickly than the sea. Warmer air rises above the land and cooler air blows in from the sea to replace it.

At night, the land loses heat more rapidly than the sea. Warmer air now rises above the sea, causing cooler air to blow out from the shore.

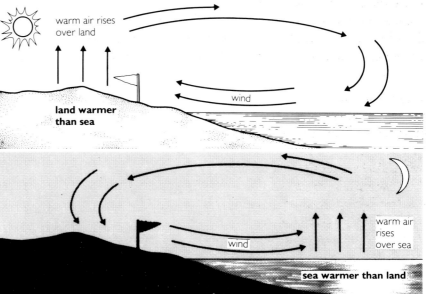

Questions

1. In a beaker containing both hot and cold water, what happens to
   a) the hot water; b) the cold water?
2. Why is a header tank fitted to a hot water system? Where is hot water from the boiler stored?
   Why is an expansion pipe necessary?
3. How does a 'radiator' distribute most of its heat?
4. Why is the freezer compartment placed near the top of a refrigerator?
5. Which heats up most rapidly during the day, land or sea? Which way do coastal breezes blow in the daytime? Why do they blow in the opposite direction at night?

Light and shade

*Finding out how light behaves is much easier than finding out what it is!
What is certain, is that light travels in straight lines and it can travel
through empty space.*

Beams and rays

In a dusty or smoky cinema it is quite easy to see the path of the
beam of light from the projector. The small particles of dust and
smoke in the air glint brightly as the light strikes them, and the edge
of the projector beam shows clearly that light travels in straight
lines.

In diagrams, lines called *rays* are used to show the direction in which
the light is travelling. You can think of a light ray as a very narrow
beam of light.

Shadows

Hold a football between a bright light and a screen and the light will
cast a shadow of the football on the screen. The shadow marks the
area which the light cannot reach because it has been stopped by the
ball.

Sharp shadow   If the light comes
from a single point such as a tiny
light bulb, the shadow of the
football is sharp and clear at the
edge.

Fuzzy shadow   If the light comes
from a spread-out source such as a
table lamp, the shadow of the
football has a fuzzy edge – around
the area of total shadow there is a
region of part shadow where only
some of the light from the lamp has
been stopped.

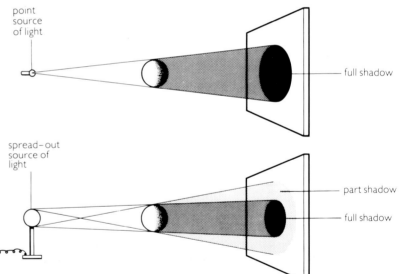

Eclipses

The Sun is a spread-out source of light and it can form gigantic
shadows of the Moon and the Earth. This gives rise to the eclipses of
the Sun and the Moon described on the opposite page. The
diagrams are not drawn to scale – it is difficult to see any detail on an
accurate scale drawing of the Sun, the Earth and the Moon:

Eclipse of the Sun

During an eclipse of the Sun, the Moon passes between the Sun and the Earth and part of the Earth's surface is thrown into shadow.

In the area of full shadow, the face of the Sun appears to be completely covered or *eclipsed* by the Moon – the eclipse of the Sun is *total*.

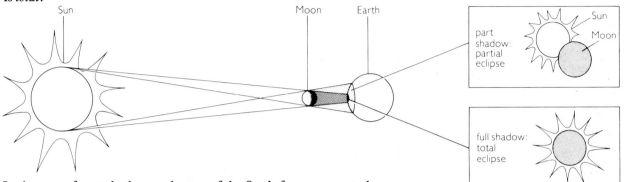

In the area of part shadow, only part of the Sun's face appears to be covered by the Moon – the eclipse of the Sun is *partial*.

The Moon circles the Earth once every 28 days but its path does not usually take it between the Sun and the Earth. For this reason an eclipse of the Sun does not happen very often.

Eclipse of the Moon

During an eclipse of the Moon, the Moon passes into the Earth's shadow – the Earth stops the Sun's rays reaching the Moon.

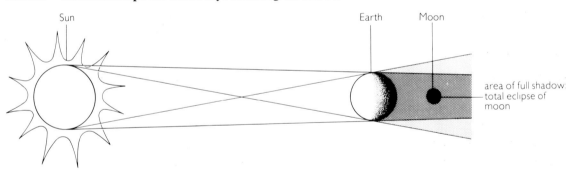

During the eclipse, the face of the Moon is not thrown into complete darkness. A little of the Sun's light still reaches the Moon by being bent through the Earth's atmosphere. In this light the Moon appears a dull coppery colour.

Questions
1. What evidence is there that light travels in straight lines?
2. What type of light source produces a sharp shadow?
3. What type of shadow does a table lamp produce?
4. Where is the Moon during an eclipse of the Sun? In a region of partial eclipse what would a photograph of the Sun look like?
5. What is the Moon's position during an eclipse of the Moon?
6. Why does an eclipse of the Sun not happen very often?

Reflections

In one side of a shiny spoon you look the right way up, in the other side you look upside down. Each side of the spoon acts like a mirror – like all mirrors, it forms an image because of the way it reflects rays of light.

The laws of reflection

A ray of light striking a mirror is reflected as shown in the diagram. The line drawn at right angles to the mirror is called a *normal* and it is from this line that the angles of the rays striking and leaving the mirror are usually measured. When a light ray is reflected:-

the angle of incidence is equal to the angle of reflection – the ray is reflected from the mirror at the same angle as it arrives;

the ray striking the mirror, the reflected ray, and the normal all lie in the same plane.

The fact that the rays 'all lie in the same plane' means that they can all be drawn on one flat piece of paper.

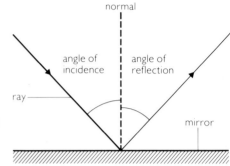

Images in flat mirrors

The diagram shows rays of light leaving a lamp and being reflected from the surface of a flat mirror. Thousands of rays leave each point of the lamp but, for simplicity, only the rays which enter the eye from a single point on the lamp have been drawn.

The rays seem to come from a position as far behind the mirror as the lamp is in front – this is where an *image* of the lamp is seen.

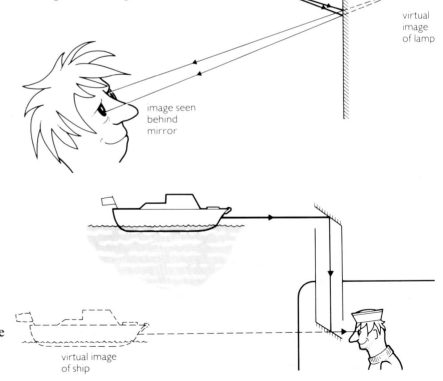

The image is described as a *virtual image* – rays seem to come from it though none actually pass through it.

The periscope

The sailor in the submarine is using an instrument called a periscope to look at a ship on the surface of the water.

In the periscope, two flat mirrors reflect light rays from the ship down to the sailor. The sailor sees an image of the ship straight ahead of him.

Convex mirrors

The surface of a convex mirror bulges outwards. Like a flat mirror, a convex mirror forms a virtual image of any object placed in front of it – but the image seen in a convex mirror is smaller than the object.

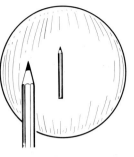

Concave mirrors

The surface of a concave mirror curves inwards. The type of image formed by the mirror depends on whether an object is placed close to the mirror or a long way from it.

Close objects   When an object is placed just in front of a concave mirror, the virtual image seen in the mirror is larger than the object. This magnifying effect makes a concave mirror very useful as a shaving or make-up mirror.

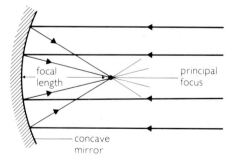

Distant objects   When rays of light from a distant object are reflected they come to a focus – they converge (come together) to form a small upside-down image that can be picked up on a screen. This image is described as a *real image* – the rays of light actually meet to form the image.

The diagram below shows a few of the rays leaving one point on a distant object. The complete image on the screen is formed by thousands of rays that leave every single point on the object.

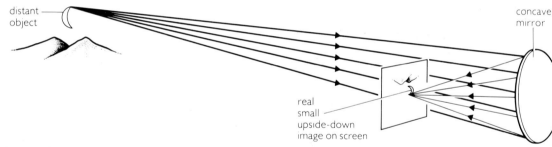

Light rays from a very distant object are very nearly parallel to each other. A concave mirror brings parallel rays of light to a focus at a point known as the *principal focus*.

The distance between a mirror and its principal focus is known as the *focal length*. Highly curved mirrors have short focal lengths.

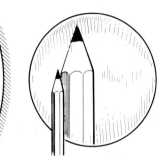

Questions

1. Draw a diagram showing a ray striking a flat mirror at an angle of incidence of 45°. What is the angle of reflection?
2. If you stand 2 metres in front of a flat mirror, where exactly is your image? Is the image real or virtual?
3. What is a periscope? How many mirrors does it contain?
4. What would your image look like in a convex mirror?
5. What would your image look like if you stood close to a large concave mirror? What is meant by:- a) the principal focus of the mirror; b) its focal length?

Bending light

In space, light travels nearly 300 000 kilometres every second – a speed which would take you nearly seven times round the world in one second. When a light ray enters glass, it slows to a speed of about 200 000 kilometres every second. As the ray slows down, it bends.

Refraction

The bending of light as it enters a transparent material such as glass or water is called *refraction*.

If a fast moving car drives at an angle into sand, it will go off course. This happens because one front wheel strikes the sand and is slowed down before the other.

A beam of light is not solid like a car but it also bends off course as it is slowed down.

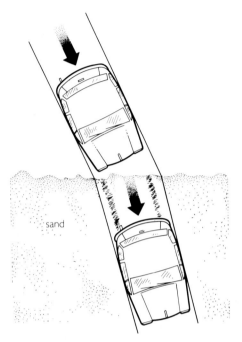

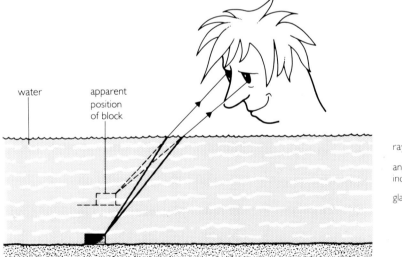

glass

light beam slows and bends

sand

Refraction in a glass block When a ray of light enters a glass block it bends *towards* the normal. As it leaves the glass block it bends *away from* the normal. If the faces of the block are parallel, the ray leaves the block parallel to the direction in which it arrived.

Apparent depth The bending of light can give you a false impression of depth.

water

apparent position of block

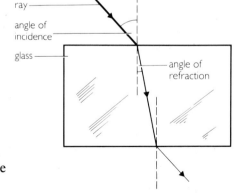

ray

angle of incidence

glass

normal

angle of refraction

Light rays from the block on the bottom of the swimming pool are bent as they leave the water surface. To the observer, the rays seem to come from a higher position – the block looks closer to the surface than it really is.

Total internal reflection

The diagram on the right shows just three of the rays leaving a lamp placed underwater.

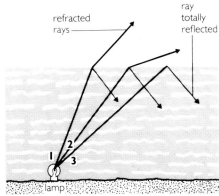

Ray 1 splits into two parts at the surface of the water. One ray passes into the air and is bent, the other weaker ray is reflected.

Ray 2 also splits into two parts. The ray passing into the air is bent so much that it only just leaves the water surface.

Ray 3 strikes the surface at too great an angle of incidence for any refraction (bending) to take place. All the light is reflected. In this case, the surface of the water acts like a perfect mirror and the ray is said to have been *totally internally reflected*.

Total internal reflection can also take place in glass. In the diagrams on the right, light rays are being totally internally reflected from inside faces of triangular glass blocks called *prisms*. Prisms can be used instead of mirrors in the periscope described on page 88.

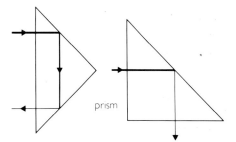

Dispersion

If a ray of white light passes through a prism as in the diagram, the light splits into a range of colours. The effect is called *dispersion* and the colour range is known as a *spectrum*.

You can see the colours of the spectrum on the rear cover of this book.

The colours are the same as those that appear in a rainbow. Most people think that they can see seven colours, but the spectrum is really a continuous change of colour from beginning to end.

This shows that white light is not one colour but a mixture of colours. It also shows that rays of different colours are refracted (bent) by different amounts.

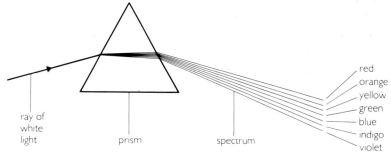

Questions

1. When a light ray passes from air into glass, what happens to its speed? Draw a diagram to show which way the ray bends.
2. Why does water appear less deep than it really is?
3. The diagram shows two rays leaving an underwater lamp. Redraw the diagram to show what happens to each ray.
4. Redraw the diagram of the periscope on page 88, replacing the two mirrors with two prisms.
5. What are the seven colours seen in a rainbow? Which colour does a prism refract (bend) the most? Which colour does a prism refract the least?

Lenses

Lenses are often made of glass and may be convex or concave in shape. Both types form images as they bend the rays of light from objects placed in front of them.

Convex lenses

The image formed when an object is a long way from a convex lens is very different from that formed when an object is close.

Distant objects   When light rays from a distant object pass through a convex lens, they *converge* (come together) to form a small upside-down image. The image is real – light rays actually pass through it and it can be picked up on a screen.

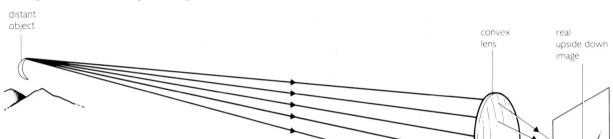

Light rays from a very distant object are very nearly parallel to each other. A convex lens brings parallel rays of light to a focus at a point known as the principal focus. The distance from this point to the lens is known as the focal length of the lens.

Finding the position of an image   If the distance of an object from a lens, and the focal length of the lens are known, the image position can be found by scale drawing.

Two rays are drawn from any point on the object:
a ray parallel to the axis – after it is bent by the lens, this ray passes through the principal focus:
a ray travelling to the centre of the lens – this ray passes through the lens without being bent.

The image is formed where the two rays meet.

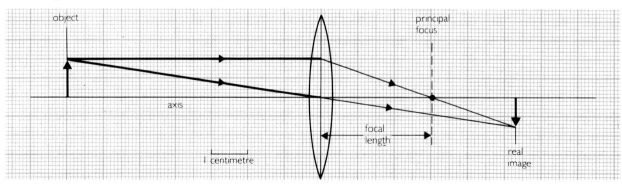

Close objects If the distance from an object to a convex lens is less than the focal length, the rays of light do not meet to form a real image:

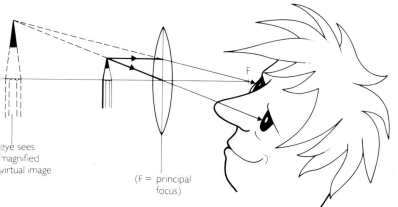

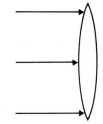

convex lens

eye sees magnified virtual image

(F = principal focus)

The diagram shows just two of the rays from one point on an object placed close to a convex lens. To the person looking through the lens, the rays *seem* to come from a position further away and a large upright image is seen. The image is virtual – no light rays actually come from there.

A convex lens used to magnify an object in this way, is often called a magnifying glass.

Concave lenses

A concave lens forms a small, upright and virtual image of any object placed in front of it – whether close or distant.

principal focus

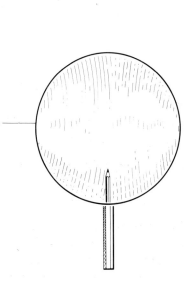

concave lens

As parallel rays of light pass through a concave lens, they diverge (spread out). The principal focus of the lens is the point from which these diverging rays appear to come.

Questions

1. What type of lens is shown in the diagram on the right? What will happen to the parallel rays of light?
2. What is the focal length of the lens in the scale diagram opposite? Copy the diagram on to a piece of graph paper. Draw the lens, the axis and the principal focus again. Draw the object 5 centimetres away from the lens. Draw in the two rays and then the image. Is the image larger or smaller than it was before? Is it closer to the lens or further away?
3. What type of lens would you use as a magnifying glass? How close must the object be to the lens?
4. What would a pencil look like through a concave lens? Is the image real or virtual?

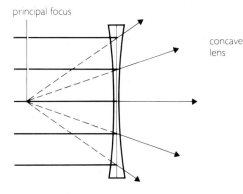

The camera and the eye

They don't exactly look alike – yet they both contain a convex lens and they both pick up images of the outside world on a rather special type of screen.

The camera

A convex lens forms a real upside down image of any distant object and this image can be picked up on a screen. In a camera, the screen is photographic film containing a light-sensitive chemical called silver bromide. Normally, the film is in total darkness. When you press the camera button, a *shutter* behind the lens opens then shuts again, quickly letting light reach the film for a brief moment only. Different amounts of light across the image cause rapid chemical changes in the film which can later be 'fixed' as a photograph.

The amount of light reaching the film is controlled by the speed of the shutter and by an adjustable ring of sliding metal plates called a *diaphragm*. The diaphragm changes the size of the hole (the *aperture*) through which the light rays pass.

The image is focused accurately by screwing the lens backwards or forwards in its holder.

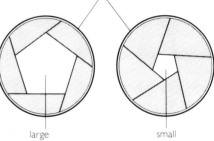

large aperture · small aperture

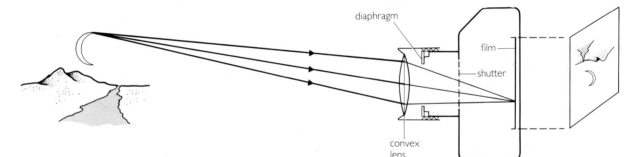

The human eye

The human eye also uses a convex lens system to form real upside-down images of objects in front of it.

The 'screen' at the back of the eye is called the *retina* and it contains nearly 130 million light-sensitive cells. These cells send nerve impulses along the *optic nerve* to the brain and the brain uses them to form a view of the outside world.

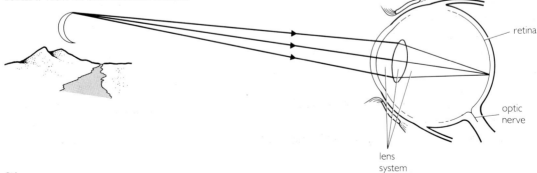

The *iris* may give you a brown, blue or hazel eye but its main job is to control the amount of light reaching the retina. If you walk into a dark room, the hole in the middle of the iris (the *pupil*) grows larger to let in more light.

Light entering the eye is brought to a focus mainly by the *cornea* and the watery liquid behind it. The *lens* itself is used to make focusing adjustments – a process called *accommodation*.

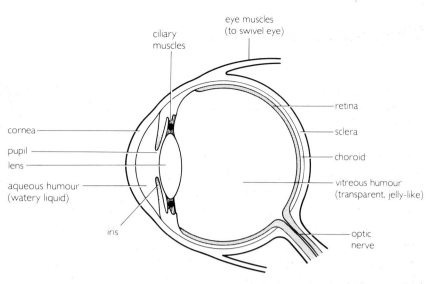

Accommodation

A perfect eye is able to form a sharp image whether an object is distant or close – it is able to *accommodate*.

The eye accommodates by changing the shape of the lens. The lens is made of an elastic substance and it thickens whenever the ring of *ciliary muscles* around it contracts. The closer an object is to the eye, the thicker the lens needs to be in order to focus light rays on the retina.

Short and long sight   Changes in the shape of the lens may not be enough to produce sharp focusing on the retina. Spectacles or contact lenses must then be worn to overcome the problem.

A *short-sighted* person cannot see distant objects clearly. The eye-ball is too long for the lens system and rays from a distant object come to a focus just in front of the retina. A *concave* lens is placed in front of the eye to correct the fault.

A *long-sighted* person cannot see close objects clearly. The eye-ball is too short for the lens system and rays from a close object try to converge just beyond the retina. A *convex* lens is placed in front of the eye to correct the fault.

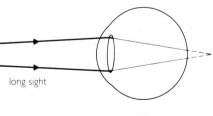

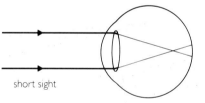

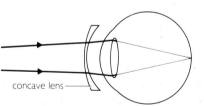

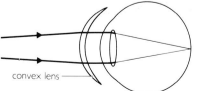

Questions

1. Where is the image formed in a camera?
2. Where is the image formed in a human eye?
3. What controls the amount of light entering a camera?
4. What controls the amount of light entering the eye?
5. How is a camera adjusted to focus light from objects at different distances?
6. What changes take place in an eye when it focuses light from objects at different distances?
7. What type of lens corrects   (a) short sight;   (b) long sight?

Colour

The eye seems to see a vast range of colours yet a TV can give you a full colour picture using three colours only.

Mixing colours

A beam of coloured light can be produced by putting a sheet of coloured plastic in front of a projector. Shine two or more beams of different colours at a white screen and a new colour is seen wherever the beams overlap.

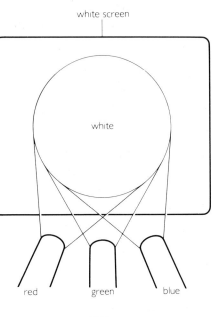

Primary colours   There are three colours that cannot be made by mixing light beams of other colours together:

these colours are *red, green and blue;*
they are known as the *primary* colours.

White light   If red, green and blue light beams overlap on a white screen, a patch of white light is seen:

red + green + blue = white

Red, green and blue together give the eye the sensation of white though pure white is made up of all the colours of the rainbow.

Secondary colours   If the red, green and blue light beams are moved apart slightly, new colours appear on the screen where any two of the beams overlap. The colours that appear are *yellow, peacock blue* and *magenta* – they are known as the *secondary* colours. You can see them on the rear cover of this book.

red + green = yellow
green + blue = peacock blue (turquoise)
red + blue = magenta (plum red)

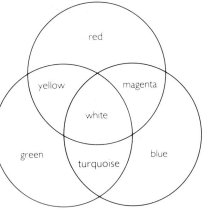

Colour television   You can give the eye the sensation of almost any colour by mixing red, green and blue light together in the right proportions. A TV screen is covered with thousands of tiny red, green and blue dots or strips which glow in different combinations to produce a full colour picture.

Seeing in colour – many colours, or three?

Most of the light-sensitive cells in the retina of a human eye do not respond to colour. Of those which do, some are sensitive to a range of colours around red, some to a range of colours around green and some to a range of colours around blue. The brain senses a colour by the effect it has on each type of colour sensitive cell – you see all colours in terms of red, green and blue.

Absorbing colour

A projector gives out its own light, so does a TV screen. Most objects are not like this – they reflect light which has come from the Sun or some other source.

Objects appear coloured because they reflect only some of the colours in the light falling on them and absorb the rest.

To the eye, white light from the Sun acts as a mixture of red, green and blue. A red dress looks red in sunlight because it reflects the red light falling on it but absorbs the green and the blue.

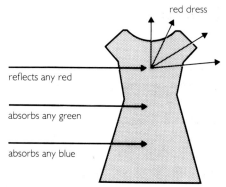

A red dress absorbs all colours except red.

A black dress appears black because it reflects no light at all. Red, green and blue light are all absorbed.

A red dress does not always look red. In the diagram on the right a red dress is being viewed in blue light:

The blue light is absorbed, there is no red light present to be reflected, so the dress appears black.

In the diagram on the right, a yellow dress is being viewed in red light:

As yellow is a mixture of red and green, the dress will reflect any red or green light falling on it. Only red light is falling on the dress, so this colour alone is reflected. The dress therefore appears red.

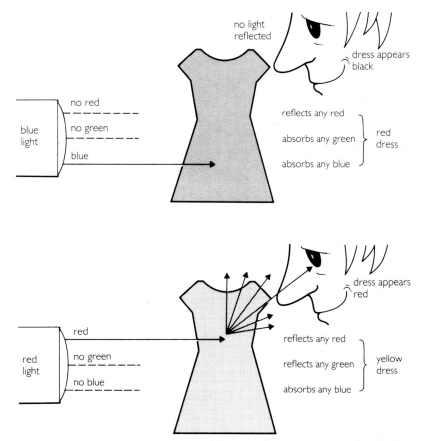

A yellow dress viewed in red light looks red.

Questions

1. What are the three primary colours? What colour is produced when these colours are added together?
2. What are the three secondary colours? What primary colours must be added to make each of them?
3. Name three objects which give out their own light.
4. Why does a black dress appear black?
5. Which colours does a red dress reflect and which does it absorb?
6. What colour will a red dress appear in green light?
7. What colour will a yellow dress appear in green light?

Electromagnetic waves

What is the difference between light of one colour and light of another?
What is light anyway?

Light waves

Light travels through space rather as ripples travel across the surface of a pond when a stone is thrown in. The ripples on a pond are disturbances which travel across the surface as water molecules slowly vibrate up and down. Light waves are disturbances which travel through space as tiny electric and magnetic forces vibrate.

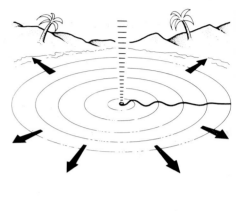

Waves in which the vibrations are from side to side or up and down are known as *transverse* waves. The ripples on a pond are transverse waves, so are light waves.

Wavelength   A series of transverse waves is shown in the diagram on the right. The distance from one wave crest to the next is known as the *wavelength*.

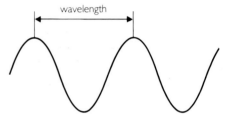

wavelength

Light waves have short wavelengths. About 1500 of the longest light waves would fit into a space only 1 millimetre long.

Colour   The eye sees light of different wavelengths as different colours. The longest light waves are seen as red and the shortest as violet. Most sources of light give out light waves of many different wavelengths.

The electromagnetic spectrum

Light waves are members of a whole family of waves known as the *electromagnetic spectrum*. The full electromagnetic spectrum is shown in the chart on the opposite page.

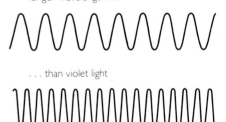

red light has a
longer wavelength . . .

. . . than violet light

Electromagnetic waves are a form of energy. They are mostly produced by electrons which are vibrating or losing energy in some way. All may be radiated from natural objects, such as the Sun; some can be generated by man-made equipment. All can travel through empty space – they do so at a speed of nearly 300 000 kilometres every second.

Electromagnetic waves differ greatly in their wavelengths:

Radio waves   have the longest wavelengths. They are used to carry sound and picture signals over long distances.

Microwaves   are radio waves of very short wavelength. Microwave beams, passed on by satellites, are used to carry telephone calls and television pictures around the world.

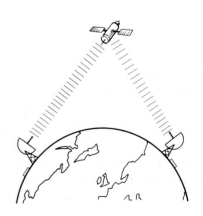

Microwaves also generate heat. A microwave oven is a very rapid means of warming food.

Infra-red rays are radiated by all warm objects (see page 100). If objects are very hot, they also radiate light – the only electromagnetic waves that can be detected by the eye.

Ultra-violet rays are present in sunlight. They are invisible to the eye but can damage the retina. It is the ultra-violet radiation from the Sun that gives you a sun tan.

X-rays and gamma rays have the shortest wavelengths in the spectrum. They have little problem in travelling through body tissue and can penetrate dense metals like lead.

Wave type		Typical uses	Typical sources
radio waves	LW MW SW	radio	radio and TV transmitters
	VHF UHF	stereo radio television	electronic circuits in which electrons are vibrated
	micro- waves	satellite communication radar microwave ovens	
infra red		electric fires	warm objects
light		electric lighting	very hot objects
ultra-violet		ultra-violet lamps sun bathing	extremely hot objects: glowing gases
X-rays		X-ray photographs	X-ray tubes
gamma rays		X-ray type photographs of metals	radioactive materials

wave length (m)

long wave

1000 — medium wave

short wave

VHF

UHF

micro waves

1/1000

infra red

1/million — light

ultra violet

1/1000 million

X-rays

1/million — gamma rays

1/million million

post office tower

red
violet

danger

radioactivity

Questions

1. Give two examples of transverse waves.
2. Draw a diagram to show what is meant by wavelength.
3. Which has the longest wavelength, red light or violet light?
4. Give one type of electromagnetic wave with a wavelength:
   (a) greater than that of light (b) less than that of light.
5. What are microwaves used for?
6. Which rays give you a sun tan? In what way are they dangerous?

Radiation

The Sun radiates vast amounts of energy in the form of electromagnetic waves. Most are infra red and light waves. They warm up anything that absorbs them and are commonly known as heat radiation – or radiation for short.

Reflectors and absorbers

On a sunny day, a car roof receives radiation from the Sun at about the same rate as a two-bar electric fire produces it. Some of this radiation is reflected by the car roof; the rest is absorbed – it heats up the inside of the car:

A black surface absorbs much more radiation than a white one – much of the radiation that strikes a white surface is reflected. In hot countries, the clothes worn are often white or lightly coloured to reflect the radiation from the Sun.

Silvery mirror-like surfaces absorb least radiation of all – they reflect nearly all of the radiation that falls on them.

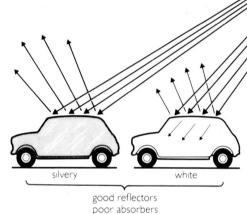

radiation

silvery   white   black

good reflectors
poor absorbers   good absorber

Radiation wavelengths

All objects *emit* (give out) some radiation. The hotter an object becomes, the more radiation it emits.

Warm objects radiate mainly long wavelength infra-red rays – they warm you up but the eye cannot detect them. As objects become hotter the wavelengths emitted become shorter. A very hot object emits some wavelengths short enough for the eye to see as red light – it glows 'red hot'.

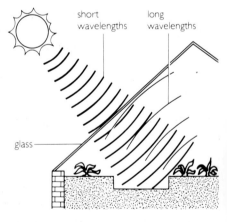

short
wavelengths   long
wavelengths

glass

Heat detectors   The photograph used on the front cover of this book is of a man drinking a cup of tea. A device which is extremely sensitive to different wavelengths in the infra-red region has recorded the radiation being given off. The device is usually used in medicine rather than tea-making, to detect ailments such as arthritis and cancer.

The greenhouse   The radiation from the hot Sun is mainly of short wavelengths. It passes easily through the glass roof of a greenhouse and warms up the materials inside. These warm materials radiate longer wavelengths that do not pass through the glass so easily. Some of this radiation is reflected back into the greenhouse – the glass 'traps' the heat.

Good and bad emitters

Some surfaces are better emitters of radiation than others. The best absorbers of radiation are also the best emitters.

Black surface are the best emitters of radiation; white surfaces are poor emitters and silvery surfaces the poorest of all. Kettles and saucepans usually have shiny outside surfaces to cut down the amount of heat wasted by radiation.

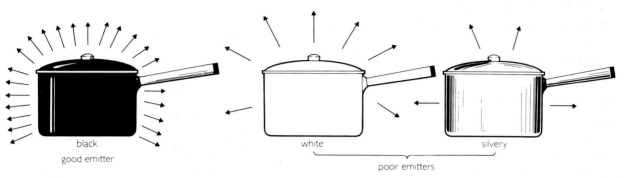

black
good emitter

white

silvery

poor emitters

Losing heat

A hot saucepan does not only lose heat by radiation. Some heat is lost by conduction – heat travels through the surrounding molecules. Some heat is lost by convection – heat is carried away by air currents rising above the saucepan.

The vacuum flask   A Thermos flask will keep liquids hot for many hours. The flask has several features designed to cut down losses of heat by conduction, convection and radiation.

The glass container has two thin walls with a narrow space between them. As most of the air has been removed from this space, very little heat can escape by conduction.

A stopper in the top of the flask stops convection currents in the air carrying heat away.

The inside surfaces of the two glass walls are silvered to reduce loss of heat by radiation.

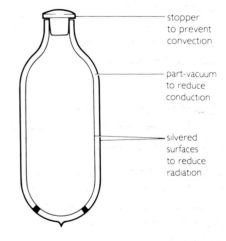

stopper
to prevent
convection

part-vacuum
to reduce
conduction

silvered
surfaces
to reduce
radiation

Questions

1. What surface is the best absorber of radiation?
2. What surface is the best reflector of radiation?
3. Why are the clothes worn in hot countries often white?
4. Give two ways in which the radiation from an object changes as the object becomes hotter.
5. Why does radiation pass more easily into a greenhouse than out of it?
6. What surface is the best emitter of radiation?
7. Why do kettles usually have a silvery mirror-like surface?
8. Give three ways in which a hot object may lose heat.
9. How are heat losses reduced in a Thermos flask?

Sound waves

Like light, sound travels in the form of waves. But sound waves are very different from light waves.

What are sound waves?

Sound waves are made when a vibrating material pushes over and over again against the substance around it.

As the cone of a loudspeaker vibrates backwards and forwards, it stretches and then squashes the air around it. These 'stretches' and 'squashes' travel outwards through the air as waves. On diagrams, the 'squashes' or compressions are often drawn as a series of lines.

As the sound waves travel outwards, they make the molecules in the air vibrate backwards and forwards. Waves in which the vibrations are backwards and forwards are known as *longitudinal* waves – sound waves are therefore longitudinal waves.

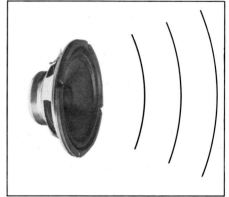

Sound waves can travel through all gases, as well as liquids and solids, but they cannot travel through a vacuum (empty space) – without a substance to be stretched and squashed, sound waves cannot be made:

The speed of sound

In air, sound waves travel at a speed of around 330 metres every second. This is about six times faster than the speed of a very fast car but slower than the speed of some aircraft. If a thunderstorm is a kilometre away, the sound of lightning takes about 3 seconds to reach you.

The speed of sound depends on the temperature of the air. Sound travels faster through warm air than through cold.

Sound travels at different speeds in different materials. In water, the speed of sound is very high – about 1400 metres every second. In most solids, the speed of sound is even higher.

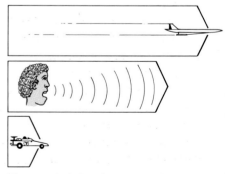

The speed of aircraft, sound, and racing cars.

Echoes

Hard surfaces such as walls reflect sound waves. When you hear an *echo*, you are hearing a reflected sound a short time after hearing the original sound:

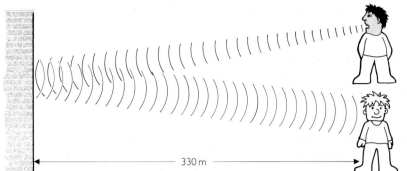

The wall on the right is 330 metres away from the person shouting. Sound waves take 1 second to reach the wall and a further second to travel back again. The second person hears the first almost immediately; he hears the echo 2 seconds later.

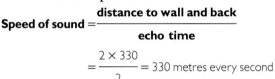

Using an echo to measure the speed of sound   If someone can measure the distance to the wall and echo time accurately, he or she can estimate the speed of sound:

$$\text{Speed of sound} = \frac{\text{distance to wall and back}}{\text{echo time}}$$

$$= \frac{2 \times 330}{2} = 330 \text{ metres every second}$$

Echo-sounding   Ships use *echo-sounding* equipment which times echoes in order to estimate the depth of water underneath them. A burst of sound sent downwards from the ship is reflected back by the sea bed. The echo time is measured – the longer the echo time, the deeper the water.

Absorbing sound waves

Echoes can be a considerable nuisance, particularly in concert halls where sounds from musical instruments may take several seconds to die down. Curtains and carpets help to absorb unwanted sounds, so too does the clothing of the audience.

Loudspeaker cabinets are often lined with sound-absorbent wadding to reduce sound reflections from the wood which might spoil the quality of the sound.

Questions

1. When does a material give out sound waves?
2. What are longitudinal waves? Give an example.
3. Why do you see a lightning flash before you hear it?
4. Does sound travel faster through hot air or through cold air?
5. Does sound travel faster through air or through water?
6. Your echo takes 1 second to reach you when you shout 160 metres away from a wall. Calculate the speed of sound.
7. What is echo-sounding equipment used to estimate? What measurement is actually taken with the equipment?
8. What materials help to cut out echoes in a room?

Inside the Albert Hall, London. Saucer-shaped sound absorbers have been hung from the ceiling to stop unwanted echoes.

103

Hearing sounds

The human ear can detect sound waves, but these sound waves may differ in several important ways.

The ear

Sound waves cause small but rapid pressure changes on anything they strike. The ear turns these pressure changes into nerve impulses which can be sensed by the brain.

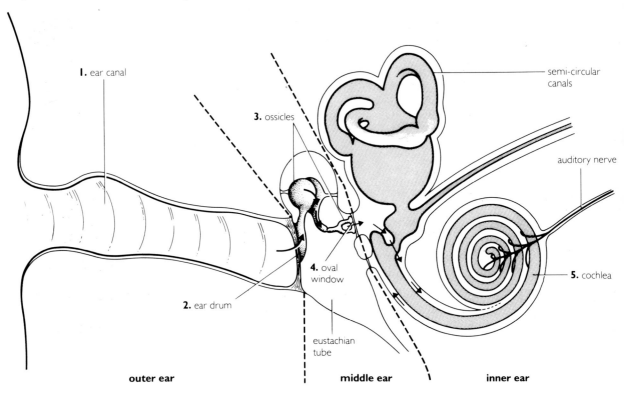

The outer ear   Sound waves passing down the *ear canal* strike a thin sheet of muscle and skin called the *ear-drum*. The sound waves make the ear-drum vibrate.

The middle ear   Behind the ear-drum lies an air-filled space containing three small bones or *ossicles*. As the ear-drum vibrates, these bones act as levers and carry the vibrations through to another sheet of skin called the *oval window*.

The inner ear   Behind the oval window is a fluid-filled tube called the *cochlea*. Vibrations travel through this fluid and are sensed by nerve endings in the tube. Nerve impulses are sent along the *auditory nerve* to the brain.

The inner ear also contains the *semi-circular canals*. These fluid-filled organs are not concerned with hearing – they give you your sense of balance.

The sound waves entering the ear can differ in several ways:

Amplitude

As sound waves pass through the air, the molecules of the air are vibrated backwards and forwards. The distance each molecule moves backwards or forwards gives the *amplitude* of the waves. The greater the amplitude of the waves, the more the air is stretched and squashed and the louder the sound seems to the ear.

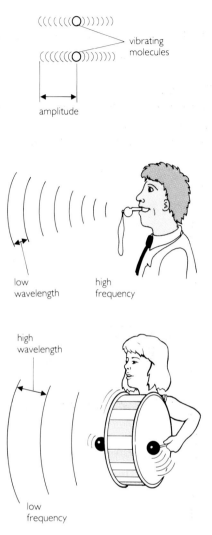

Frequency

A bass drum gives out about 20 sound waves every second; a whistle might produce about 10 000 sound waves every second. The number of sound waves produced every second is known as the *frequency*.

Frequency is measured in *hertz* (Hz). If a loudspeaker cone vibrates 110 times every second, 110 sound waves are given out every second – the frequency of the sound waves is 110 Hz.

Wavelength

The distance between one compression and the next is known as the *wavelength*. The more sound waves are produced every second, the closer together the compressions will be – the higher the frequency of the waves, the lower their wavelength.

The speed, frequency and wavelength of the waves are linked by an equation:

speed of sound = frequency × wavelength

Since the speed of sound is 330 metres every second, a loudspeaker of frequency 110 Hz will produce sound waves of wavelength 3 metres:

330 = 110 × 3
m/s   Hz   m

Questions

1. In the ear, how are sound vibrations transferred from the ear-drum to the oval window? Which part of the ear changes sound vibrations into nerve impulses?
2. Which part of the ear is concerned with sense of balance?
3. If the amplitude of the sound waves from a loudspeaker were increased, what change would you notice in the sound?
4. The sound waves from a loudspeaker have a frequency of 1000 Hz. What does this mean? If the frequency were increased, how would this affect the wavelength of the waves?
5. What equation connects speed, frequency and wavelength? Sound waves from a loudspeaker have a frequency of 160 Hz and a wavelength of 2 metres. What is the speed of sound?

The sounds of music

high frequency
high pitch

low frequency
low pitch

To the ear, some sounds seem higher than others. High or low, the sound of a guitar seems very different from the sound of a piano.

Frequency and pitch

The human ear can detect sound waves with frequencies ranging from 20 Hz up to about 20 000 Hz, though the ability to hear the higher frequencies becomes less with age.

Different frequencies sound different to the ear. High frequency sound waves are heard as high notes – the *pitch* is said to be high; low frequencies are heard as low notes – the pitch is low.

Octaves   Musical instruments normally produce notes based on octaves. Two notes are an octave apart in pitch if one has twice the frequency of the other.

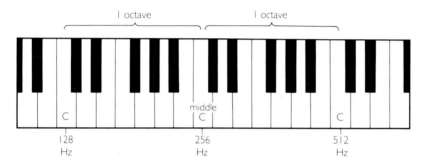

The piano keyboard shown above is tuned to *scientific pitch*. To overcome certain tuning difficulties, pianos are normally tuned to frequencies slightly different from those shown.

Vibrating strings

Instruments like the piano and the guitar give out sound waves when stretched wires or strings are made to vibrate.

A guitar string produces sound waves when it is plucked. The frequency, and therefore the pitch, of the note will rise if the string is tightened or the vibrating length of the string is shortened. A guitar string is effectively shortened when you push it against one of the frets with your finger:

If the vibrating length of a guitar string is halved, the frequency of the note doubles – its pitch increases by one octave.

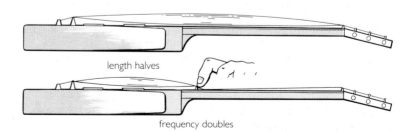

length halves

frequency doubles

Amplitude   The amplitude of a vibrating string is the greatest distance it moves sideways as it vibrates – the larger the amplitude, the louder the sound.

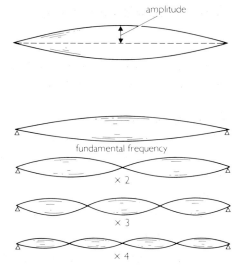

An electric guitar has a *volume* control to increase the amplitude of the sound waves given out by the loudspeakers.

Harmonics   The diagram on the right shows the simplest way in which a guitar string can vibrate. The frequency of vibration is known as the *fundamental frequency*. The guitar string could however vibrate at twice or three times this frequency, or more.

These various frequencies of vibration are known as *harmonics* and some or all may be present in a vibrating string at the same time. It is the mixture of harmonics present that gives a musical instrument its own characteristic sound – a guitar sounds like a guitar and not like a piano.

Resonance in pipes

Like a guitar string, a column of air has frequencies at which it will naturally vibrate provided there is something to set off the vibrations.

The air in a milk bottle starts to vibrate if you blow gently across the top of the bottle. The sound grows loud if the blown air vibrates at a frequency that exactly matches the natural frequency of vibration of the air in the bottle – the effect is called *resonance*.

Wind instruments like the recorder produce sounds when resonance takes place in a pipe containing air. The air vibrations which cause resonance in a recorder come from a flow of air that breaks up as you blow across a narrow wedge.

Long pipes produce lower notes than short pipes   Resonance in a long pipe takes place at a lower fundamental frequency than in a short pipe. To produce different notes on a recorder you change the pipe length by blocking air holes with your fingers.

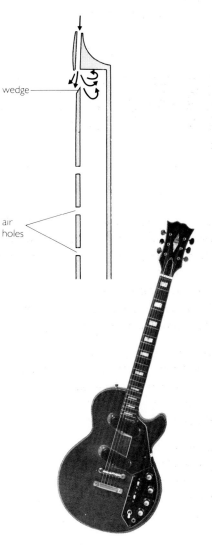

Questions

1. If the frequency of a sound increases, does the sound seem higher or lower?
2. A sound has a frequency of 200 Hz. What is the frequency of a sound   (a) 1 octave higher;   (b) 2 octaves higher?
3. What happens to the pitch of a guitar string if you   (a) halve the length of string which is able to vibrate;   (b) slacken the string?
4. What is meant by the amplitude of a vibrating string?
5. What makes a guitar sound like a guitar and not a piano?
6. When resonance takes place, which produces the higher notes, a long pipe or a short pipe?

At which fret would you press to increase the pitch of the strings by one octave?

107

Further questions

1 How are the movements of molecules in a gas different from those in a solid?

2 What happens to the molecules in a material when the temperature falls?

3 $-373\,°C, -273\,°C, -20\,°C, 0\,°C, 23\,°C, 37\,°C,$ $90\,°C, 100\,°C, 200\,°C$
From the above list, select:
    a the temperature at which water freezes
    b the temperature at which water boils
    c absolute zero
    d the temperature on a warm sunny day
    e the average temperature of the human body

4 Drop ink into water and the colouring soon spreads throughout the water.
Explain why this happens.
What is the process called?

5 If, through a microscope, you were to observe tiny particles of smoke floating in air, what would you see and why?

6 Why do small amounts of water try to form into round droplets?

7 Why do materials such as cotton wool and blotting paper 'soak up' water?

8 How is 'rising damp' prevented in the walls and floors of a modern house?

9 Most materials expand when heated.
    a When is expansion a nuisance?
    b What use can be made of expansion?
    Give one example in each case.

10 The diagram below shows a bimetal strip.
    a Draw another diagram to show what will happen to the bimetal strip when it is heated.
    b Name two devices which make use of a bimetal strip.

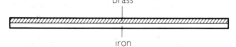

11 In what way is the expansion of water different from that of most other substances?

12 *Copper, iron, glass, water, air*
In the above list:
    a Which is the best conductor of heat?
    b Which is the poorest conductor of heat?

13 The metal end of a spade feels cold to touch, yet the wooden handle feels warm. Explain.

14 **a** Copy the diagram above and mark the position at which the viewer will see an image of the cross.
    b On your diagram, draw two rays from the cross to show how the image is formed.

15 **a** What types of lenses are shown above?
    b A matchstick is held upright close to lens **A**. How would the matchstick appear when viewed through the lens?
    c How would the matchstick appear if lens **B** were used instead of lens **A**?
    d If the matchstick were moved a long way back from lens **B**, how would this affect its appearance when viewed through the lens?

16 A carpet appears green when viewed in white light.
    a What colour(s) does it reflect?
    b What colour(s) does it absorb?
    c What colour would it appear if viewed in red light?

17 Which of the following waves are not electromagnetic?
radio, light, sound, X-rays, infra red

18 Heat energy may be transferred from one place to another by conduction and convection. What is the third possible method of heat transfer? How does heat energy reach us from the Sun?

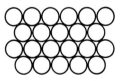

19 **a** Use a ruler to measure the wavelengths of the two waves **A** and **B** shown above.
b Both waves travel at the same speed. If the frequency of **A** is 2 Hz, what is **B**'s frequency?

20 One sound has a high pitch, another has a low pitch. What is the difference between the two sets of sound waves?

21 Give two ways in which the frequency of a vibrating guitar string could be increased.

22 This diagram shows how the atoms of a certain substance are thought to be arranged.

a How can you tell from the diagram that the substance is a solid?
b When the substance is heated, it expands.
i What happens to the atoms as the substance gets warmer? **ii** How does this help to explain why the substance expands?
c If the heating is continued, the substance eventually melts. What difference will this make to the arrangement of the atoms? SEG/SWEB

23 **a** The clinical thermometer shown below uses the fact that liquids expand when heated.

i Name a suitable liquid for L. Give a reason for your answer. **ii** What is the purpose of the part labelled P? **iii** Explain how a nurse would reset the thermometer for use.
iv What is special about the temperature 37 °C? **v** Suggest reasons why the thermometer scale only measures between 35 °C and 43 °C.
b A metal screw top has become jammed on top of a glass jar.

i Explain how you would release the top.
ii Explain why this particular method works. LEAG

24 When water in a pond freezes, ice first forms on the water surface.
i Why does this happen? **ii** Why is this of importance to living things? SEG

25 This diagram shows the inside of a refrigerator.

insulation

i Why are the walls of the refrigerator insulated? **ii** Name a suitable insulating material that could be used. **iii** Explain why the cooling unit, X, is placed at the top. LEAG

26

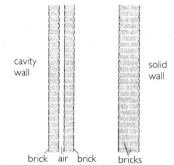

cavity wall

solid wall

brick air brick

bricks

a Why do air-filled cavity walls keep a house warmer in winter than solid brick walls?
b Why does filling the cavity with plastic foam keep the house even warmer?
c Explain how a hot water radiator heats a room.

MEG

27 The diagram below shows the main parts of a solar heating system designed to provide hot water for a house. Heat energy from the sun warms the water in the solar panel. This water is then pumped through copper tube inside the hot water tank so that it can transfer its heat to the water in the tank.

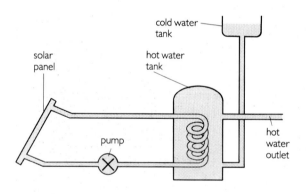

cold water tank

solar panel

hot water tank

pump

hot water outlet

a Why is copper a good material from which to make the spiral?
b State one reason why the tube is bent into a spiral rather than being straight.
c The pipes between the solar panel and the hot water tank are lagged.
i Name a suitable material for the lagging.
ii State one reason why this should be done.
d Explain why the hot water outlet pipe is at the top of the tank.

SEG

28 a Copy and complete the diagram to show what happens to light rays entering the eye in the case of short sight.

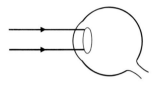

b What type of lens would be used to correct this defect?
c This diagram shows the pupil in daylight.

Redraw the diagram to show how the pupil would be if a person was in a darkened room.

SEG

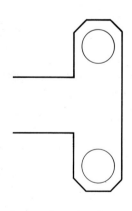

29 a Copy and complete the diagram of a simple camera. Show and label the following parts: film, lens, shutter, aperture.
b The inside of the camera is painted black. Why is this?
c The camera is used to take two pictures
1. A dimly lit close up
2. A bright distant view
i What is moved to refocus the camera?
ii In which direction is this moved for the second picture?   **iii** How can the camera be adjusted to let in the right amount of light for the second picture? Give two methods.

SEG/SWEB

Chapter 4

Forces and energy

What makes a car stop when it runs out of petrol?

...... is being used to create a .... that accelerates the dragster ...

Speed and acceleration

Some cars go faster than others.
Some cars gain speed faster than others.

Speed

The police can check the speed of a car with a radar 'gun'. But there's a simpler method. Measure the distance between two points along a road, say two lamp posts. Measure the time a car takes to travel between these points. Then calculate the speed:

$$\text{Average speed} = \frac{\text{distance moved}}{\text{time taken}}$$

For example, if a car travels 80 metres in 4 seconds:

$$\text{Average speed} = \frac{80}{4}$$

$$= 20 \text{ metres per second}$$

This can be written 20 m/s for short.

On most journeys the speed of a car changes, so the actual speed isn't always the same as the average speed. To find an actual speed, you have to find the distance travelled in the shortest time you can measure.

Speed can be measured with a radar gun.

Acceleration

From a standing start, a racing car can reach a speed of 50 m/s in 10 s or less. It gains speed very quickly. It has a high acceleration.

To work out the acceleration, you use the equation:

$$\text{Acceleration} = \frac{\text{gain in speed}}{\text{time taken}}$$

For example, if a racing car gains 50 m/s of speed in 10 seconds:

$$\text{Acceleration} = \frac{50}{10}$$

$$= 5 \text{ m/s per second}$$

This tells you that the speed of the car is going up by 5 m/s every second. This acceleration can be written 5 m/s^2, for short.

Retardation is the opposite of acceleration. If a car has a retardation of 5 m/s^2, its speed goes *down* by 5 m/s every second.

Grand Prix cars accelerating.

Graphs

Graphs are a useful way of showing how something is moving.

A rally car is travelling along a road at a steady speed of 20 m/s. The co-driver starts her stopwatch as the car passes a lamp post. She checks the speed every second and writes it down. These are her results:

time in seconds	0	1	2	3	4
speed in m/s	20	20	20	20	20

A second rally car is stopped by the lamp post. When the co-driver starts his stopwatch, the car accelerates away at 4 m/s². This is how the speed changes:

time in seconds	0	1	2	3	4
speed in m/s	0	5	10	15	20

Questions

1. One of the rally cars above is leaking oil. It drips one drop of oil on the road every second. Can you tell which rally car it is from the diagram?
    Draw a diagram to show how the drips would look if the other car were leaking oil instead.
2. A car is travelling along a motorway at a steady speed. It travels 150 m in 5 seconds:
    (a) What is its average speed?
    (b) How far does it travel in 1 second?
    (c) How far does it travel in 6 seconds?
    (d) How long does it take to travel 240 m?
3. Copy and complete:
    A motor cycle has a steady . . . . . . of 3 m/s². This means that every . . . . . . its . . . . . . increases by . . . . . .
4. The graph on the right shows how the speed of a car changes.
    (a) What is the maximum speed of the car?
    (b) For how many seconds does the car stay at its maximum speed?
    (c) How much speed does the car gain in the first 20 seconds? How much speed does it gain every second? What is its acceleration?
    (d) What happens to the car after 40 seconds?

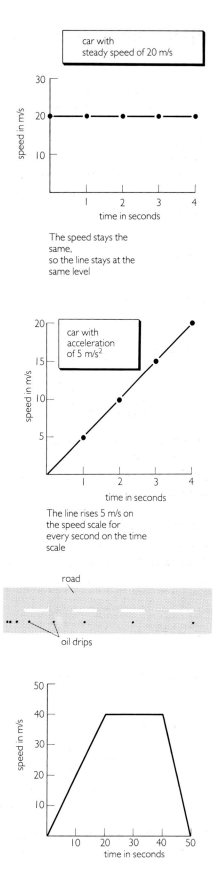

car with steady speed of 20 m/s

The speed stays the same, so the line stays at the same level

car with acceleration of 5 m/s²

The line rises 5 m/s on the speed scale for every second on the time scale

road

oil drips

Force, mass and weight

Most people think they know what a force is – but what is the difference between mass and weight? This section explains this but they would never believe you in the local supermarket.

Force

Force is an amount of push or pull and it is measured in *newtons* (N for short).

As forces go, a force of 1 newton (1 N) is very small. The astronaut in the diagram will use a force of about 20 newtons to pull the ring off the top of her can of coke.

The rocket motor in her spacecraft will produce a force of about 50 000 newtons.

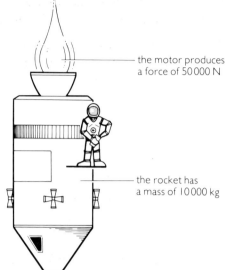

the motor produces a force of 50 000 N

the rocket has a mass of 10 000 kg

Mass

Mass is the quantity of matter in an object and it is measured in *kilograms* (kg for short).

The can of coke has a mass of less than ½ kilogram.
The astronaut has a mass of about 70 kilograms.
Her spacecraft might have a mass of about 10 000 kilograms.
The mass of the Earth is a staggering
6 000 000 000 000 000 000 000 000 kilograms!

How masses behave   All objects have mass. This affects the way they behave in two ways:

Large masses are more difficult to speed up or slow down than small masses.

If the astronaut pushes her can of coke it speeds up easily – the same would not be true if she tried to push her spacecraft. The behaviour of moving masses is described in the next section.

All masses attract each other

No one has yet discovered why.

The *mass* of the sugar is one kilogram.

Gravitational force

The force of attraction between masses is known as *gravitational force*. The larger and closer the masses, the greater is the pull between them.

The gravitational force between the Earth and the Sun holds the Earth in orbit around the Sun; the gravitational force between the Moon and the Earth holds the Moon in orbit around the Earth.

The gravitational force between two everyday objects is too weak to measure but it becomes quite strong if one of the objects has as much mass as the Earth – strong enough for example to hold the astronaut firmly on the Earth's surface. If she drops her can of coke, gravitational force will quickly pull it towards the Earth.

Weight

Weight is the common name for the gravitational force on an object and it is measured in newtons like all other forces.

On Earth, the astronaut has a weight of 700 newtons – that is the strength of the Earth's gravitational pull on her 70 kilograms of mass.

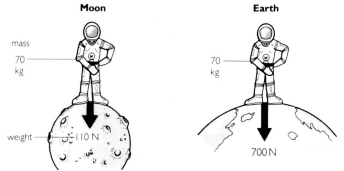

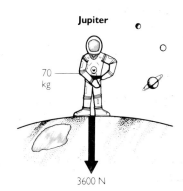

If the astronaut stands on the Moon, her weight will be much less because the Moon has less to attract her with. On a massive planet like Jupiter the astronaut would weigh more than five times as much as she does on Earth.

The astronaut's weight may vary but she still has the same amount of matter in her. Wherever she stands, her mass will still be 70 kilograms.

Measuring weight and mass

If you put a one-kilogram bag of sugar on a set of scales, it is the gravitational pull on the sugar which forces the scale pan downwards – the scales respond to weight.

Some sets of scales are marked in newtons but most giving a reading in mass units such as kilograms or grams (there are 1000 grams in each kilogram). Readings in mass units present no problem to manufacturers of scales because an object of large mass also has a large weight on Earth.

As for the one-kilogram bag of sugar:
it has a *mass* of 1 kilogram and a *weight* on Earth of 10 newtons, but try telling them that at the supermarket!

The *weight* of the sugar is 9.8 newtons.

Questions

1. In what units are forces measured?
2. In what units are masses measured?
3. Give two ways in which large masses differ from small masses.
4. What is the common name for gravitational force?
5. Why does an astronaut weigh less on the Moon than the Earth? How is her mass affected if she moves from the Earth to the Moon?
6. How many grams are there in 1 kilogram?
7. Why is it wrong to say that a bag of sugar 'weighs' 1 kilogram? What is the weight of the sugar on Earth?

Force and movement

Deep in space, spacecraft can move with no forces on them at all. But on Earth, moving vehicles have friction to slow them down.

Moving without force

If something is moving, and there are no forces on it, it keeps on moving – at a steady speed in a straight line. This is what happens to a spacecraft deep in space. Far from the pull of gravity, the spacecraft keeps moving for ever – even with its engine turned off.

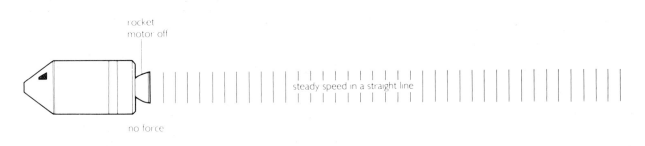

The effect of a force

If the spacecraft fires its rocket engine, it will accelerate. The *greater* the force from the engine, the *greater* the acceleration.

With the *same* force on it, a spacecraft with *more* mass would have *less* acceleration.

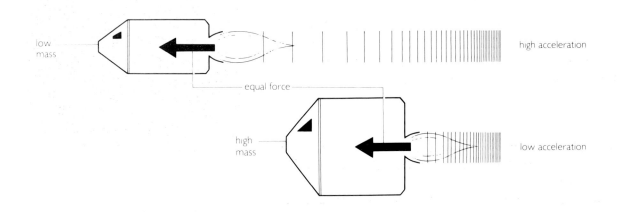

There is an equation linking force, mass and acceleration:

Force = mass × acceleration
　　N　　　kg　　　m/s²

For example:
A 1 newton force would make a 1 kg mass accelerate at 1 m/s².
A 6 newton force would make a 3 kg mass accelerate at 2 m/s².

Friction

An unpowered craft may move at a steady speed in space but on Earth a moving car will begin to slow down the instant its engine is switched off. It is slowed by friction, the force which exists between all materials which rub against each other.

Air resistance is one type of frictional force. The resisting force from the air flowing past a fast moving car may be 2000 newtons or more.

In the diagram on the right, another type of frictional force is acting – that between two solid surfaces pressed together. The car body is difficult to move because of friction between its bottom surface and the ground.

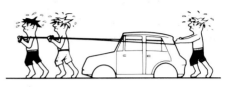

Overcoming friction   People solved the problem of friction from the ground by putting vehicles on wheels. Even a wheel produces some friction where it rubs against its shaft. Oil reduces the friction by helping the surfaces slide across each other more smoothly.

To reduce friction still further, wheels on vehicles are usually mounted on ball or roller bearings.

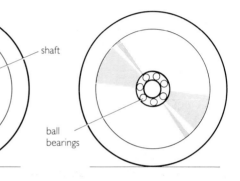

Using friction   Friction does have its uses. If there were no friction between car tyres and the ground, the wheels of the car might spin but the car would stay where it was. Car tyres are designed to give as much friction as possible in all weather conditions.

Friction is also used to slow moving vehicles. In a disc brake, a steel disc attached to the car wheel spins between two small brake pads. When the brakes are applied the pads press against the disc and the frictional force slows the disc down.

Questions

1. What happens to a moving spacecraft with no force on it?
2. What happens to the spacecraft on the right if it fires its rocket motor? How would the result be different if the spacecraft had more mass?
3. What force is needed to make a 10 kg mass accelerate at 3 m/s²?
4. A spacecraft has a mass of 1000 kg. Its engine gives a force of 4000 N. What is its acceleration? (There are no other forces on the spacecraft.)
5. Why does a fast car slow down if its engine is switched off?
6. Give two ways in which the friction between a wheel and a shaft can be reduced.
7. Where on a wheel is as much friction as possible needed? Why? Give one other use of friction.

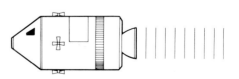

Forces for rockets and jets

You can't make a force unless there's something to push against.
But what does a rocket push against in space?

Reaction forces

The astronaut and the rock shown below are 'floating' together in space. When the astronaut pushes the rock with his boot, the rock moves to the left and he moves to the right – he cannot push on the rock without it pushing equally hard on him.

When any force acts there is an opposite reaction force of the same strength.

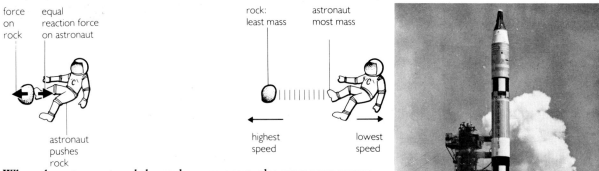

| | | | | |
| force on rock | equal reaction force on astronaut | | rock: least mass | astronaut most mass |

astronaut pushes rock

highest speed   lowest speed

When the astronaut and the rock move apart, the astronaut moves more slowly than the rock – he has the greater mass and is not so easily moved by a force. To move at a higher speed, the astronaut would either have to push harder on the rock or push against a more massive rock.

The rocket motor

In pushing against the rock the astronaut has made himself into a simple rocket motor.

Rocket motors don't normally push rocks, they throw out a large mass of fast moving gas. In pushing the gas out, a rocket motor is pushed in the opposite direction by the reaction force.

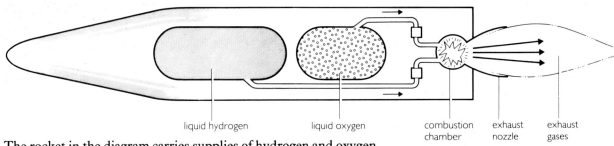

liquid hydrogen   liquid oxygen   combustion chamber   exhaust nozzle   exhaust gases

The rocket in the diagram carries supplies of hydrogen and oxygen stored as cold liquids so that they take up less room. In the combustion chamber the hydrogen burns violently with the oxygen and the gases formed expand rapidly in the intense heat. The expansion forces the gases out of the exhaust nozzle at high speed.

Jet engines

Like rocket motors, jet engines produce a forward force by pushing out large masses of gas behind them. Unlike rocket motors they will not work in space because they need a supply of air.

Why air is needed   A jet engine needs air for two reasons.

First, air provides mass for the engine to push out. In large quantities air is heavy stuff – there are probably more than 100 kilograms of air just in your sitting room at home.

Second, air provides the oxygen needed to burn the kerosene fuel. Rockets have to carry their own bulky supply of oxygen with them.

How the engine works   Large amounts of air are drawn in to the engine by the spinning blades of the compressor – the compressor looks like a series of gigantic fans:

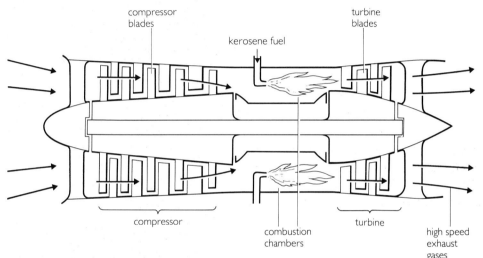

The compressor pushes air into the combustion chambers where kerosene (a type of paraffin) is squirted in. The kerosene burns fiercely in the air and the hot gases which form are forced out of the exhaust nozzle at high speed.

Before leaving the engine the gases pass through sets of turbine blades. The turbine blades are pushed round by the gas and the rotation is used to drive the compressor.

Questions

1. If the astronaut in the diagram on the opposite page pushes the rock with a 30 newton force, what reaction force acts on him? Why does he move more slowly than the rock?
2. Why are rocket fuels stored in liquid form?
3. A jet engine needs air for two reasons. What are they?
4. Describe what happens to the air as it passes through a jet engine. What drives the compressor round?
5. Give one advantage of a rocket motor over a jet engine.
6. Give one advantage of a jet engine over a rocket motor.

Forces which turn

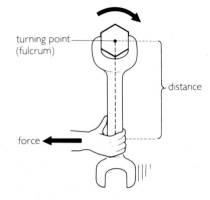

How far can a bus be tilted before it falls over? What makes it fall over anyway? This section looks at the way forces makes things turn.

The turning effect of a force

The spanner in the diagram is being used to tighten a bolt. The force of the hand on the handle is producing a turning effect – the spanner and the bolt turn about a point known as a *turning point* or *fulcrum*.

There are two ways of producing a stronger turning effect, as shown by the diagrams in the margin. You can either pull with a greater force, or pull at the end of a longer handle. If you want it really tight, you might use both!

Moments   The strength of the turning effect produced by a force is called a *moment*. It can be calculated as follows:

$$\text{Moment of a force} = \text{force} \times \text{shortest distance from the force to the turning point}$$

If the force increases, or its distance from the turning point increases, its moment also increases.

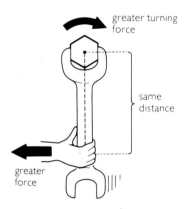

Moments in balance

The plank in the diagram is balanced on a log of wood. The plank was balanced first, then bricks were placed on in positions that did not disturb the balance. Each of the bricks used weighs 20 newtons. Each set of bricks produces a turning effect – one moment tries to tip the plank to the left, the other tries to tip it to the right. The plank balances because both moments are equal:

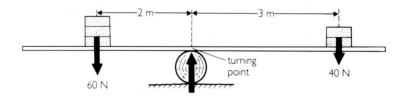

Moment (turning to the left) = Moment (turning to the right)

Force	×	distance from turning point	=	Force	×	distance from turning point
60 N	×	2 m	=	40 N	×	3 m
120 Nm			=	120 Nm		

In each case, the turning moment is 120 Nm.

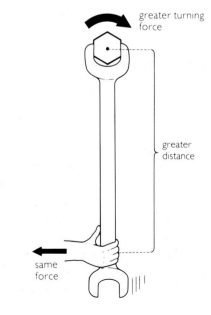

Centre of gravity

Lift a plank and it acts as if all its weight were concentrated at one point only. The point is known as the centre of gravity.

Support a plank at its centre of gravity and it will balance. Try to lift it at any other point and the turning effect of its weight will make it tip.

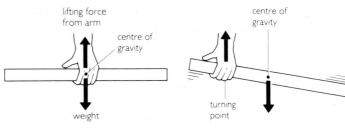

Stability

Some things topple over much more easily than others – it all depends on the position of the centre of gravity.

Push a box a little then release it and it will fall back to its original position – the box is in a *stable* position.

Push the box too far and it will topple. It starts to topple as soon as its centre of gravity has passed over the edge of its base – beyond this point its weight has a turning effect which tips the box even further.

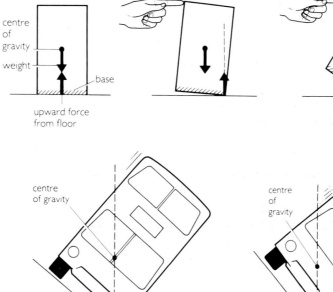

Factors affecting stability A bus has a wide base. It has a low centre of gravity because its heavy mechanical parts are low down. These factors make it very stable – the bus has to be tilted a long way before it topples:

A coach is even more stable than a double-decker bus. It has a lower centre of gravity and can be tilted further before it topples.

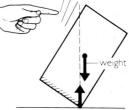

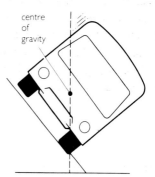

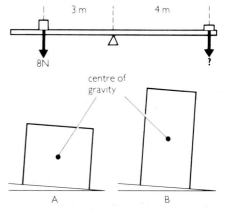

Questions:

1. How can you produce a strong turning effect on a bolt?
2. What is a *moment*? How is the moment of a force calculated?
3. What weight is needed to balance the beam in the diagram?
4. Why is it easiest to carry a plank by holding it at its centre of gravity?
5. Which of the two boxes on the right is the most stable? Which two factors make it the most stable? Draw a diagram of each box tilted to the point at which it is about to topple.

Work and energy

When is a jumbo jet like 5000 chocolate biscuits? Which releases most energy – eating a chocolate biscuit, throwing it, or dropping it?

Work

You do work when you climb the stairs, or lift a suitcase, or kick a football. Cranes do work when they lift a load and jet engines do work as they push aircraft through the air.

Work is done whenever a force moves.

1 joule (J) of work is done when a force of 1 newton moves through a distance of 1 metre.

The work done when a force moves can be calculated using the equation:

Work = force × distance

A 3 N force moving 4 m does 12 J of work.

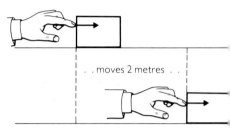

When a 3 newton force . . .

. . moves 2 metres . .

. . . 6 joules of work are done

Energy

Things have energy if they can do work.
Like work, energy is measured in *joules*.

There are many different forms of energy, though energy can change from one form into another.

Kinetic energy   Throw a stone and you give it energy. The moving stone has energy because it can do work – it does work if it smashes a window and it does work if it knocks something over. In doing work, the stone loses energy and slows down.

The energy of the moving stone is known as *kinetic energy*. The faster the stone is moving, the more kinetic energy it has. If two stones are thrown at the same speed, the stone with the greater mass has more kinetic energy.

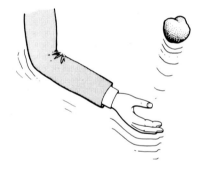

Potential energy   An object has potential energy if it has been pushed or pulled into a position from which it can do work. A compressed spring has potential energy, so does a stretched rubber band – both can do work when released.

Lift a stone above the ground and you give it potential energy. Work was done in lifting the stone and the stone will do work when it falls – by crushing anything underneath it. The heavier and higher the stone, the more potential energy it has.

Heat energy   Heat is a form of energy because it can be used to do work. Burning petrol is used to turn a car engine.

The heat energy put into a material often makes the molecules move faster – all the molecules gain extra kinetic energy.

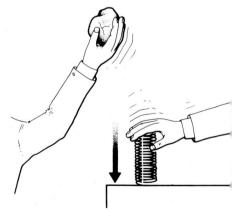

Chemical energy   You need energy to throw a stone. The energy comes from food and it is released in your body when the food is combined with oxygen. The energy stored in food is called chemical energy.

Fuels like petrol and coal also contain chemical energy. When fuels burn in oxygen, chemical energy is changed into heat energy.

Electrical energy   The electric motor in the diagram is using electrical energy to do work. The energy carried to the motor by the electric current was originally stored in the battery as chemical energy.

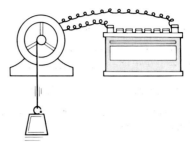

Other forms of energy   The list above is by no means complete. Light and radio waves carry energy, so do sound waves. Nuclear energy is stored in the nucleus of each atom.

Some typical energy values $\quad$ 1000 joules = 1 kilojoule (kJ) and 1 000 000 joules = 1 megajoule (MJ)	
The work you might do in lifting a building brick up above your head from the ground level	50 joules (50 J)
The kinetic energy the brick might have if you throw it	75 joules (75 J)
The work you might need to do to climb stairs	1 500 joules (1.5 kJ)
The energy you gain by eating a chocolate biscuit	300 000 joules (300 kJ)
The energy gained by drinking a pint of beer	600 000 joules (600 kJ)
The energy needed to boil a kettleful of water	700 000 joules (700 kJ)
The energy you gain from a large helping of chips	1 000 000 joules (1 MJ)
The energy stored in a car battery	2 000 000 joules (2 MJ)
The energy gained from all the food you eat in one day	11 000 000 joules (11 MJ)
The energy released by burning a gallon of petrol	160 000 000 joules (160 MJ)
The kinetic energy of a jumbo jet as it takes off	1 500 000 000 joules (1500 MJ)

Questions
1. When is work done? How much work would you do if you pushed a lawnmower 3 metres with a force of 20 newtons?
2. Give an example of something which has kinetic energy.
3. Give two examples of things which have potential energy.
4. What type of energy is stored in   (a) food   (b) fuels?
5. What device uses electrical energy to do work?
6. Some of the energy values in the chart can vary greatly. Which values will depend on:
   (a) how heavy you are   (b) how tall you are
   (c) how strong you are   (d) how fast something is moving?
   Which are examples of chemical energy?

Heat energy

A jumbo jet can't get energy from chocolate biscuits but many other energy changes are possible. Energy often changes from one form to another, but it usually ends up as heat energy.

Energy changes

When anything happens anywhere in the universe, energy changes from one form into another. Here are two simple examples:

A stone falling to the ground As a stone falls, it loses height and gains speed – its potential energy is changed into kinetic energy.

When the stone hits the ground, both the stone and the ground heat up slightly. The stone stops moving but the molecules in the stone and the ground move a little faster – the kinetic energy of the stone has been changed into heat energy.

A stone thrown across the ground As the man throws the stone, chemical energy in his muscles is changed into kinetic energy.

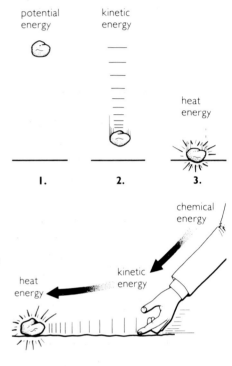

The stone skids across the ground, generating heat as it is slowed down by friction – its kinetic energy is changed into heat energy. In both of the examples above:

the total amount of energy at the end of each process is the same as it was at the beginning;

all the energy eventually ends up as heat energy.

These two statements remain true for all energy changes. All the different types of energy that there are, eventually turn into heat energy.

Absorbing heat energy

When a stone absorbs heat energy, its temperature rises.

The stones in the diagram are being heated up to the same temperature; both are absorbing heat from bunsen flames. More heat energy is needed to raise the temperature of the larger stone – there are many more molecules in it to be speeded up.

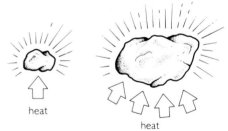

With more heat energy put into it, the larger stone has more heat energy to get rid of when it cools down. When the bunsens are turned off, the larger stone will continue to give off heat long after the smaller stone has cooled down.

Some materials are much better at storing heat than others:

a 1 kilogram stone made of granite needs to absorb only 840 joules of heat energy for its temperature to rise by 1 °C;
1 kilogram of water needs to absorb five times as much heat for its temperature to rise by the same amount.

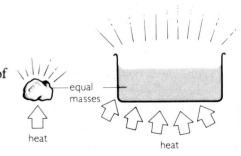

The specific heat capacity of a substance is the heat energy needed to raise the temperature of 1 kilogram of the substance by 1 °C. It is measured in joules per kilogram per °C [J/(kg °C)].

The table below gives the specific heat capacities of some common substances:

Specific heat capacities: J/(kg °C)

Hot water bottles – good for storing heat.

Energy in joules needed to heat 1 kilogram through 1 °C					
Water	4200	Concrete	850	Steel	500
Sea water	3900	Granite	840	Copper	400
Meths	2500	Aluminium	900	Glass	670

Water has a particularly high specific heat capacity – it takes 4200 joules of energy to heat 1 kilogram of water through 1 °C. One kilogram of water will give out 4200 joules of heat energy if it cools down by 1 °C. This makes water a useful substance for storing or carrying heat energy – examples are shown on the right.

Concrete has a lower specific heat capacity than water but a higher density – in the same space there is more mass. Concrete blocks are used in night storage heaters to store heat energy. Electrical heating elements heat the blocks overnight when electricity is cheaper to buy. The hot blocks continue to release their heat energy through the day as they slowly cool down.

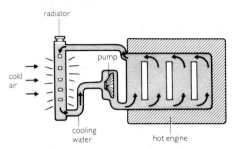

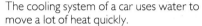

The cooling system of a car uses water to move a lot of heat quickly.

Calculating heat energy changes

The heat energy gained or lost as the temperature of a substance rises or falls can be calculated using the equation:

$$\frac{\textbf{Heat}}{\textbf{energy}} = \textbf{mass} \times \frac{\textbf{specific heat}}{\textbf{capacity}} \times \frac{\textbf{temperature}}{\textbf{change}}$$

For example, it would take 5400 joules of heat energy to increase the temperature of 3 kilograms of aluminium by 2 °C:

$$\begin{array}{ccccccc} 5400 & = & 3 & \times & 900 & \times & 2 \\ \text{J} & & \text{kg} & & \text{J/(kg°C)} & & \text{°C} \end{array}$$

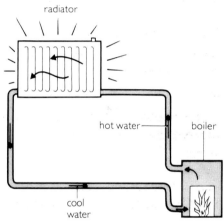

Central heating systems use water to spread heat around the home.

Questions

1. A falling stone hits the ground. What happens to its kinetic energy? What happens to its temperature?
2. Water has a high specific heat capacity. When can this be useful? Give examples.
3. Using the same heater, which would heat up most rapidly, a kilogram of sea water or a kilogram of ordinary tap water?
4. Copper has a specific heat capacity of 400 J/(kg °C). What does this mean? Calculate how much heat energy is needed to raise the temperature of 3 kilograms of copper by 4 °C.

Latent heat

When the stones in the previous section lost heat energy, their temperature fell. Yet a pan of hot wax may lose heat for an hour or more without getting cooler. Something is happening to the wax even if its temperature isn't changing . . .

When a pan of hot liquid wax is removed from the heat, it starts to cool. If its temperature is measured every minute or so, a graph can be plotted as in the diagram:

The temperature of the liquid wax falls until the fat starts to turn solid. The temperature then stays unchanged for an hour or more. It does not start to fall again until the liquid wax has fully solidified.

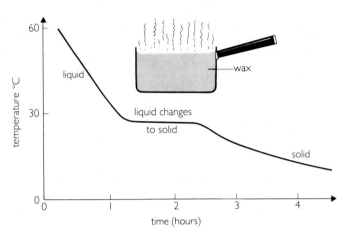

Latent heat of fusion

The wax gives off heat energy as it solidifies yet it does not get any cooler. There seems to be some hidden supply of heat energy inside it. The molecules themselves are the source of this hidden or *latent* heat. They lose some of their freedom and energy as they bind together in solid form, and it is this energy which is released as heat.

If the solid wax is heated it soon starts to melt. Its temperature stays unchanged as it melts. All the heat energy absorbed by the wax is used in giving the molecules enough energy to make them change from the solid to the liquid state.

The heat energy needed to change a solid into a liquid is known as the latent heat of fusion.

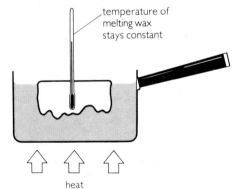

Latent heat of vaporisation

A liquid must be supplied with heat energy if it is to change into a gas.

The heat energy needed to change a liquid into a gas is known as the latent heat of vaporisation.

Water is turned into a gas every time a kettle is boiled. It takes nearly seven times as much heat energy to change a kilogram of water into steam as it does to change a kilogram of ice into water.

Evaporation

When a liquid changes into a gas, it is said to *evaporate*. Heating speeds up evaporation but a liquid evaporates whether it is heated or not. Wet clothes dry and rain puddles disappear even on a cold day because fast-moving molecules of water are constantly escaping from the liquid to form gas. Other than by direct heating, there are several ways of making a liquid evaporate more rapidly:

Increase the surface area of the liquid Wet roads dry out quickly because the rain-water is spread over a large area. This gives the molecules a greater chance of escaping from the liquid.

Pass a stream of air through a liquid or across its surface A liquid evaporates more rapidly if air is bubbled through it because it then has a larger surface area in contact with the air.
Air flowing across a liquid surface also helps evaporation: wet washing dries more quickly on a windy day.

Form the liquid into a fine spray A spray is made up of millions of tiny liquid droplets and the total surface area of these droplets is very large. The liquid molecules escape easily from the highly curved surface of each droplet.

In the carburettor of a car engine, petrol evaporates rapidly as it is drawn into the airstream in the form of a fine spray. The petrol gas is sucked into the engine with the air and then burnt.

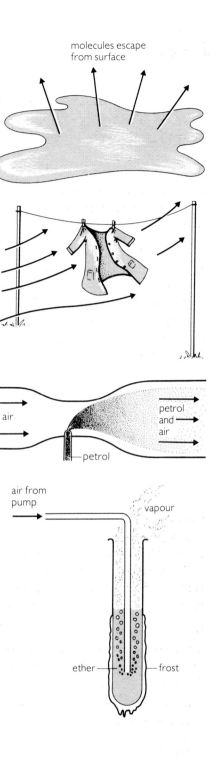

Cooling by evaporation

A liquid needs heat from somewhere in order to evaporate. If the liquid is not actually being heated, it must draw the heat energy it needs from some other source.

A little methylated spirit tipped on the back of your hand quickly evaporates. It extracts the latent heat that it needs from your hand, which feels very cold as a result.

If air is bubbled through ether as in the diagram, the latent heat needed to evaporate the either is drawn from the glass test-tube. As the ether evaporates, the glass may get so cold that frost begins to form on it. (*Warning:* experiments with ether have to be carried out under very carefully controlled conditions.)

Questions
1. What happens to the temperature of hot wax as it cools?
2. Why is heat energy needed to melt a solid? What is this heat energy called?
3. What is latent heat of vaporisation?
4. Why do wet roads dry so quickly after a shower of rain?
5. Petrol enters the airstream through a carburettor in the form of a fine spray. Why?
6. Why does your hand feel very cold when you tip a little methylated spirit over it?

More about evaporation

When a liquid evaporates, it draws heat from everything it touches. A refrigerator uses the cooling effect of evaporation, so does the human body – but it's more difficult to keep the body cool on a muggy day.

The refrigerator

The cooling effect in many refrigerators is produced when a liquid called Freon evaporates. Freon is a very *volatile* liquid – it evaporates easily.

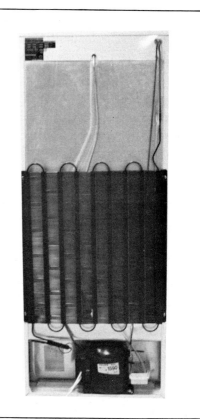

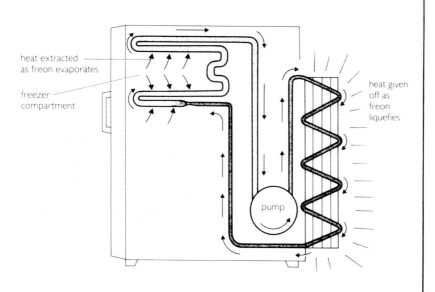

heat extracted as freon evaporates

freezer compartment

heat given off as freon liquefies

pump

The liquid Freon evaporates in the network of tubes in the freezer compartment. It does so as more and more Freon gas is drawn away by the electric pump. As the Freon evaporates, it draws the latent heat it needs from the food and the air next to the tubes – so everything in the freezer compartment cools down.

The pump compresses the Freon gas and forces it into the zig-zag pipe at the back of the refrigerator. Here the gas *condenses* – it changes back to liquid – and releases its latent heat. The unwanted heat is lost through the cooling fins around the pipe. The pipe feels hot if you touch it. The Freon, now liquid, is pumped round to the freezer tubes where it evaporates again.

Cooling the body

The human body produces a great deal of unwanted heat. It must get rid of this heat in order to stay at a steady temperature of about 37 °C. Much of the heat is lost by convection but some is lost as moisture on the skin evaporates. Moisture diffuses through the skin all the time. As it evaporates, it draws the latent heat it needs from the body which is cooled as a result.

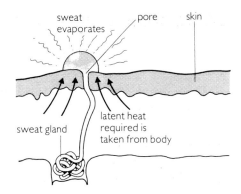

Sweating You start to sweat if your blood temperature rises more than about ½ °C above normal. The sweat is largely made up of water and it comes from *sweat glands* just under the surface of the skin. The sweat emerges through tiny holes in the skin called *pores*. As it evaporates, it takes heat away from the body.

Sweating also starts if the temperature outside your body rises above about 30 °C.

Humidity

When air is humid, it is heavily laden with water in gas form – the gas is called *water vapour*. Water evaporates slowly in humid air because molecules from the water vapour condense back into the water almost as fast as those in the water escape into the air.

Sweat tends to stay on your skin on humid or 'close' days. You feel hot and uncomfortable because the sweat cannot evaporate. A breeze helps to cool you because it aids evaporation.

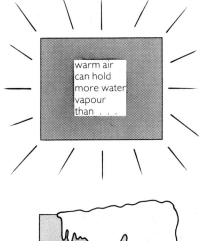

Saturated air Air is said to be *saturated* when it is holding the maximum possible amount of water vapour.

Warm air can hold more water vapour than cold air. If warm saturated air is cooled, some of its water vapour must condense back to liquid form. The water may appear as dew on the ground, or millions of tiny droplets in the air – we see these water droplets as mist or clouds.

Cold surfaces cool the air which touches them and water vapour may condense as a result. Condensation on cold windows and cold water pipes is caused in this way. When condensation freezes, it forms frost.

Questions

1. What do the tubes around the freezer compartment of a refrigerator contain? What makes these tubes so cold?
2. A refrigerator gives off heat energy. Where? *Why* is heat energy released?
3. When does the human body start to sweat? Where does the sweat come from?
4. Why do you lose heat when you sweat?
5. Why does a breeze help to keep you cool?
6. What is humid air?
7. Why does condensation form when warm, wet air is cooled?

Four-stroke engines

They change chemical energy into kinetic energy. They use up valuable resources. But life wouldn't be the same without them.

The four-stroke petrol engine

The diagram below shows a single cylinder petrol engine in action. The *crankshaft* of the engine is pushed round every time the piston is forced down in the cylinder by the explosion of an air and petrol mixture. For each explosion that takes place there are four down and up movements of the piston – the engine moves through a *four-stroke cycle:*

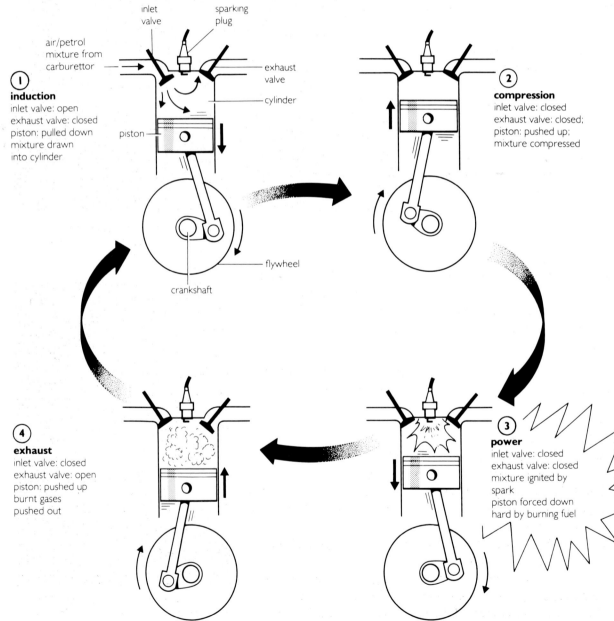

1 induction
inlet valve: open
exhaust valve: closed
piston: pulled down
mixture drawn
into cylinder

2 compression
inlet valve: closed
exhaust valve: closed;
piston: pushed up;
mixture compressed

3 power
inlet valve: closed
exhaust valve: closed
mixture ignited by
spark
piston forced down
hard by burning fuel

4 exhaust
inlet valve: closed
exhaust valve: open
piston: pushed up
burnt gases
pushed out

The carburettor supplies the cylinder with air and petrol mixed in the correct proportions for an explosion.

The sparking plug ignites the mixture with a high voltage spark produced by an *ignition coil*.

The flywheel gains enough kinetic energy during power strokes to keep the engine turning through the other strokes.

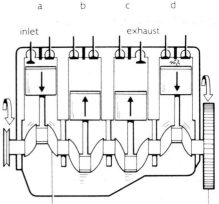

Four-cylinder engines Most car engines contain four cylinders with all the pistons turning the same crankshaft. Each cylinder is on a different stroke. This means that there is always one cylinder on a power stroke, so the crankshaft is pushed round more smoothly.

A *distributor* is used to send sparks from the ignition coil to each of the four plugs in turn.

Wasted energy When petrol burns in the cylinders of an engine, energy is released. Only about a quarter of this energy actually reaches the wheels of the car. The rest is wasted as heat. Most of this waste heat is lost with the exhaust gases. But some has to be carried away by the engine's cooling system.

General view of a car engine. The carburettor, spark plugs, ignition coil and distributor are all visible.

Diesel engines

Diesel engines also follow a four-stroke cycle. But they use fuel oil instead of petrol, and they don't need spark plugs. Air entering the engine is compressed so much that it becomes very hot. The fuel oil is squirted straight into the cylinders. The oil ignites as soon as it meets the hot air.

Diesel engines are heavier and more expensive than petrol engines. But they use less fuel, and they wear out more slowly.

Questions

1. What does a carburettor do?
2. Why are engines fitted with flywheels?
3. Most car engines have four cylinders or more. Why?
4. Study the diagram of the four-cylinder engine near the top of the page. Each cylinder is at the beginning of a stroke. Write down which stroke each is on.
5. During which stroke is energy released from the petrol? What happens to most of this energy?
6. Write down two ways in which diesel engines are the same as petrol engines, and two ways in which they are different.
7. What are the advantages of diesel engines over petrol engines?

A cut-away view of the same engine. How many of its components can you identify?

Machines

Pulleys, scissors and gearboxes are machines – so are human arms. You never get more work out of a machine than you put into it, but it does enable the work to be done more conveniently.

Levers

The lever shown below is a simple machine which makes it easier for the man to lift the load. The man's downward force, or *effort*, has twice as much turning effect on the lever as the load, so an effort of 40 newtons is all that is needed to lift the 80 newton load up off the ground.

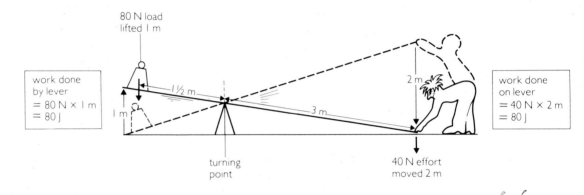

Mechanical advantage   Comparing the load moved with the effort required gives the *mechanical advantage* of a machine:

$$\text{Mechanical advantage of a machine} = \frac{\text{load}}{\text{effort}}$$

The lever shown above has a mechanical advantage of two. The lever would have an even higher mechanical advantage if the load were nearer the turning point or the effort further away.

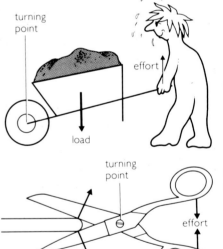

The wheelbarrow and the scissors on the right are both machines with a mechanical advantage greater than one – in other words, a large load is overcome with a small effort. But there is a price to pay for this gain in force: the effort has to move further than the load.

In the human arm, the load lifted at the end of the arm is much smaller than the effort produced by the muscle. The arm may lose in lifting ability but it gains in moving ability – only a small movement of the muscle is needed to move the end of the arm through a large distance.

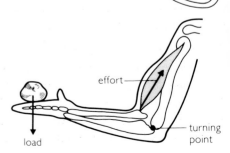

Doing work   The work done by any force can be calculated by multiplying the force by the distance it moves. In the diagram above, the load is twice as large as the effort but it only moves half as far. As the calculations on the diagram show:

The work done by the lever = the work done on the lever

Pulleys

The pulley on the right is being used to lift up an engine – pulley wheels are usually mounted side by side in the pulley blocks but it is easier to draw them one above the other.

The mechanic pulls with an effort of 100 newtons to lift the 300 newton load: the pulley has a mechanical advantage of three.

Velocity ratio For the load to be lifted 1 metre, each of the numbered ropes has to shorten by 1 metre. Each rope is of course part of the one long rope, so the mechanic has to pull this rope down a total of 4 metres.

This means that the effort moves four times further than the load – the pulley has a *velocity ratio* of four.

$$\textbf{Velocity ratio of a machine} = \frac{\textbf{distance moved by effort}}{\textbf{distance moved by load}}$$

Efficiency Most machines waste energy because of friction, so more work has to be put into a machine than is got out. The *efficiency* of a machine can be worked out as follows:

$$\textbf{Efficiency} = \frac{\textbf{work done by a machine}}{\textbf{work done on a machine}}$$

The mechanic in the diagram does 400 joules of work on the pulley rope; the pulley only does 300 joules of work on the load. The pulley is ¾ efficient – only ¾ of the work done on the pulley is actually used to lift the load; the rest is wasted.

The efficiency of a machine can be worked out in another way:

$$\textbf{Efficiency} = \frac{\textbf{mechanical advantage}}{\textbf{velocity ratio}}$$

The pulley in the diagram has a mechanical advantage of three and a velocity ratio of four – figures again giving an efficiency of ¾.

If a machine has an efficiency of 1, it does not waste any energy. Its mechanical advantage (comparing forces used) is equal to its velocity ratio (comparing distances moved).

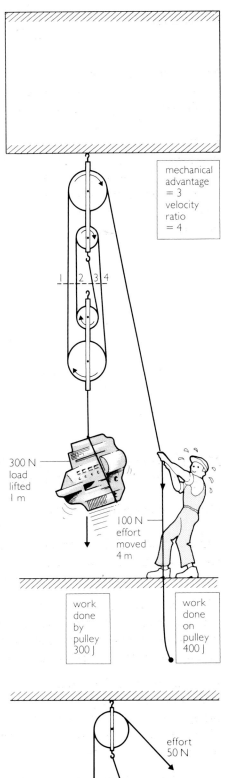

mechanical advantage = 3 velocity ratio = 4

300 N load lifted 1 m

100 N effort moved 4 m

work done by pulley 300 J

work done on pulley 400 J

effort 50 N

load

Questions

1. How is the mechanical advantage of a machine calculated?
2. Name two machines with a mechanical advantage greater than 1.
3. Name a machine in which the effort is greater than the load.

The remaining questions refer to the pulley on the right:

4. If the load is to be lifted 1 metre, how far must the rope be pulled? What is the velocity ratio of the pulley?
5. The pulley has an efficiency of one. What does this mean?
6. What is the mechanical advantage of the pulley? What load can the pulley lift with an effort of 50 newtons?

The human engine

A petrol engine burns petrol, a human engine 'burns' food. Like other engines, the human body can do work when it releases the energy stored in its fuel.

Food fuels

Nearly all the energy obtained by the human body comes from three types of food – *carbohydrates*, *fats* and *proteins*.

Carbohydrates   These come from foods such as bread, flour and other cereals, sugar, and potatoes. They provide you with about 50% of your energy.

carbohydrate

Carbohydrates are compounds of carbon, hydrogen and oxygen which break down when digested into substances like *glucose* (a simple sugar).

Glucose is carried round the body in the bloodstream. Some is changed into *glycogen*, which is stored in the liver and muscles and can be converted back to glucose whenever its energy is needed. Energy is released when glucose is combined with oxygen in the cells of the body:

glucose + oxygen → carbon dioxide + water + energy

This is a process of slow combustion without the explosions and flames found in a petrol engine but it does provide the energy needed to move the muscles of the body. We breathe in air to get the oxygen needed for combustion, and we breathe out carbon dioxide and water vapour as our exhaust gases.

If the body takes in more carbohydrate than it needs to produce energy, the surplus is converted into body fat.

Fats   Main sources of fats are foods such as butter, meat, milk and margarine. The fats you eat supply you with about 40% of your energy.

Like carbohydrates, fats are made up of carbon, hydrogen and oxygen and they release energy when combined with oxygen.

As the body can store fats, they act as a useful reserve supply of energy. Nonetheless, most doctors think that people would be less liable to heart and bowel disease if the amount of fat in their diet was reduced.

Proteins   Animals are largely made of substances called proteins, and plants are partly made of them.

The body needs protein for the growth and repair of body tissues but any spare protein is used as fuel.

You obtain most of your protein from foods such as bread, milk, eggs, cheese, meat, and fish.

Energy from food

The table on the right shows the amounts of energy that can be obtained from a number of different foods. Fats are the most concentrated source of energy.

Energy needs   The energy needs of average 16 year-olds are:

females – about 9 500 kilojoules (kJ) every day;
males – about 12 500 kilojoules (kJ) every day.
(1 kilojoule = 1000 joules)

If you are 16 and female, 3 kilograms of boiled potatoes would probably provide you with all the energy you need in one day – though they would not give you all the other essential nutrients necessary for good health.

Energy use   About half the energy you gain from your food is used to drive essential processes in the body – energy is needed for breathing, circulating blood and maintaining body temperature. The rest of the energy gained from food is needed so that your muscles can do work.

When you use up energy to move your muscles, only about 15% of it is converted into work. The other 85% is released as heat energy – which is why physical exercise makes you sweat.

energy stored in 100 grams of food (about ¼ lb)	
lard	3700 kilojoules
sugar	1600 kilojoules
corn flakes	1500 kilojoules
bread	1000 kilojoules
boiled potatoes	330 kilojoules
celery	35 kilojoules
(1 kilojoule = 1000 joules)	

Sources of food energy

All the food which you eat comes from plants, or animals which feed on plants.

Green plants use the Sun's energy to build up carbohydrates, fats and proteins from carbon dioxide, water, and minerals in the soil – the process is called *photosynthesis*.

The Sun is therefore the source of all energy stored in your food.

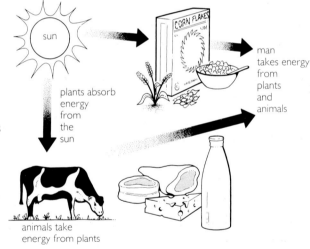

Questions

1. Which types of food supply us with most of our energy?
2. What substance is formed when carbohydrates are digested? Where can the body store this substance and in what form?
3. Energy is released when glucose is combined with oxygen. What else is produced?
4. Which types of food are the most concentrated source of energy?
5. Which stores most energy – 100 grams of cornflakes or 100 grams of lard?
6. Why do you sweat when you do physical work?
7. Where does all the energy in food originally come from?

What type of food is each of these?

Energy for the world

Industrial countries need huge amounts of energy. Most comes from fuels which are burnt in factories, homes, vehicles and power stations. But supplies of these fuels are running out.

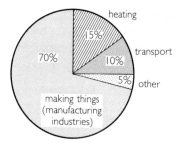

How energy is used in Britain

Many power stations use coal, oil or natural gas as fuel. The heat from the burning fuel is used to make steam. Jets of steam turn huge 'fans' called turbines. These turn the generators which make electricity.

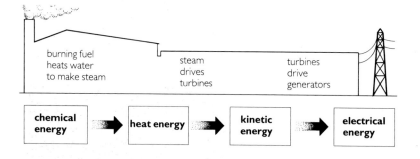

Renewable and non-renewable energy sources

In many countries, wood is the main fuel used for cooking and heating. Wood is a **renewable** source of energy. When used, more can be grown to replace it.

Coal, oil and natural gas are **non-renewable** energy sources. They can't be replaced, and the Earth's supply of them is gradually running out. At present rates of use:
there is enough oil and natural gas left to last for 50–100 years
there is enough coal left to last for 200–300 years.
So, the search is on for new sources of energy.

Alternative energy sources

Here are some alternatives to coal, oil and natural gas. Most are renewable. Once used, the energy is replaced naturally.

Wind energy   Giant windmills are used to turn electrical generators.
For   Renewable energy source. Doesn't cause pollution.
Against   Windmills are large and costly, with relatively low power output. Not all areas have enough wind to drive them.

Hydroelectric energy   A dam is built across a river. A lake forms behind the dam. Fast-flowing water from the lake is used to turn generators.
For   Renewable energy source. Doesn't cause pollution.
Against   Few areas of the world are suitable. New lakes change the landscape and upset the balance of animal and plant life.

A wind-driven generator.

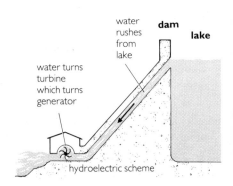

136

Tidal energy   A dam is built across an estuary. A lake behind the dam fills up at high tide and empties at low tide. Fast flowing water turns generators.
For   Renewable energy source. Doesn't cause pollution.
Against   Very expensive to set up. Few areas are suitable.

Solar energy   Mirrors and panels are used to capture the Sun's radiant energy – usually as heat.
For   Renewable energy soure. Doesn't cause pollution.
Against   Continuous sunshine needed.

Nuclear energy   Radioactive materials naturally release heat. A nuclear reactor speeds up the process. The heat is used to generate electricity (see page 242).
For   Small amounts of nuclear fuel give large amounts of energy.
Against   Radioactive materials give out dangerous radiation. Waste materials from nuclear power stations stay radioactive for hundreds of years.

Geothermal energy   Water is heated by the hot rocks which lie many kilometres beneath the Earth's surface. The hot water is used to make steam to turn generators.
For   Renewable energy source. Huge quantities of energy are available. Doesn't cause pollution.
Against   Deep drilling is very difficult and expensive.

Biomass energy   Fast growing plants, called biomass, are used to make alcohol. The alcohol is used as a fuel, like petrol.
For   Renewable energy source.
Against   Huge land areas needed to grow plants. This may upset the balance of nature.

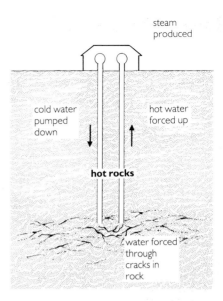

Filling up with alcohol.

Questions

1. (a) What is the difference between a *renewable* and a *non-renewable* source of energy?
   (b) Which of these are renewable sources of energy?
      coal   oil   biomass   wind   tides   natural gas
2. biomass   hydroelectric   solar   wind   nuclear   geothermal

   Which of the above words would you pick to describe each of the following?
   (a) energy radiated by the Sun
   (b) energy in hot rocks
   (c) energy from radioactive materials
   (d) energy from fast-flowing water
   (e) energy from fast-growing plants
3. On the right is a block diagram of a hydroelectric scheme. Each box is a type of energy. Copy the diagram and complete the boxes to show the energy changes which take place.

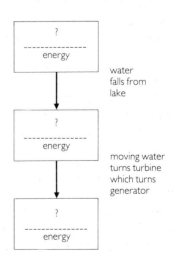

Density

Which weighs more, a tonne of coal or a tonne of feathers? Which weighs more, a cubic metre of coal or a cubic metre of feathers?

Everybody knows that feathers are 'lighter' than coal . . . but are they?

The pile of feathers in the rather unlikely experiment on the right is just as heavy as the pile of coal.

The feathers are of course taking up far more space than the coal – they have a larger volume. A fairer test might be to compare *equal volumes* of coal and feathers.

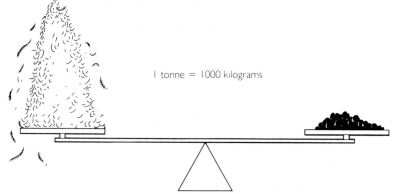

1 tonne = 1000 kilograms

Density

All the substances shown in the chart below have the same volume. Each box has a volume of 1 cubic metre and the mass of material that fills it is marked in kilograms (kg).

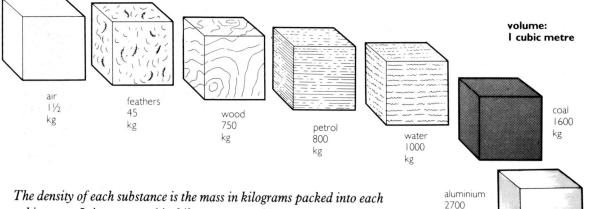

volume: 1 cubic metre

air 1½ kg
feathers 45 kg
wood 750 kg
petrol 800 kg
water 1000 kg
coal 1600 kg
aluminium 2700 kg
steel 7800 kg
lead 11300 kg

The density of each substance is the mass in kilograms packed into each cubic metre. It is measured in kilograms per cubic metre – sometimes written kg/m^3 for short.

Some examples from the chart:

The density of lead	11 300 kilograms per cubic metre.
The density of coal	1600 kilograms per cubic metre.
The density of water	1000 kilograms per cubic metre.
The density of the feathers	45 kilograms per cubic metre.

So feathers are not 'lighter' than coal, they are *less dense*.

Other density units Density values are sometimes given in grams per cubic centimetre (g/cm^3). Water for example has a density of one gram per cubic centimetre – one cubic centimetre of water has a mass of one gram.

Measuring density

Unfortunately, materials are not usually found in exact cubic metre chunks. Densities have to be measured using more awkward shapes and volumes.

The density of a substance can be worked out once the mass and volume are known:

If 6 cubic metres of water have a mass of 6000 kilograms,
1 cubic metre of water has a mass of 1000 kilograms.
The density of the water is 1000 kilograms per cubic metre – the figure is worked out by dividing the number of kilograms by the number of cubic metres. So:

$$\textbf{Density} = \frac{\textbf{mass}}{\textbf{volume}}$$

Measuring mass   The mass of a substance can be found using a mass balance or a set of scales marked in kilograms (or grams).

Measuring volume   The volume of a liquid can be found by pouring the liquid into a measuring cylinder and reading the level on the scale.

The volume of a small solid can be found by lowering the solid into a measuring cylinder partly filled with a liquid such as water. When the solid is fully immersed, the rise in liquid level on the scale gives the volume of the solid.

If the solid is in the shape of a simple block with square corners, its volume can be calculated:

Volume = length × width × depth

Relative density

Sulphur has a density of 2000 kg per cubic metre; water has a density of 1000 kg per cubic metre. Sulphur is therefore twice as dense as water – it has a relative density of 2.

$$\frac{\textbf{Relative density}}{\textbf{of a substance}} = \frac{\textbf{density of substance}}{\textbf{density of water}}$$

Water itself has a relative density of 1.

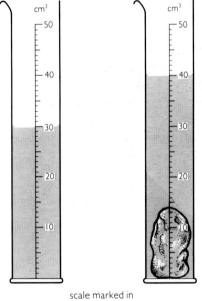

scale marked in
cubic centimetres

Questions

1. What does the density tell you about a substance?
2. Write down the densities of the following substances: lead, coal, air, petrol, aluminium, water, steel.
3. What is the volume of the liquid in the measuring cylinder on this page? What is the volume of the small solid placed in it?
4. Sulphur has a relative density of 2. What does this mean?
5. In the diagram on the right, which substance has the greatest mass – A, B, C, or D? Which has the greatest volume? Work out the density of each substance. Use the chart on the opposite page to help you decide which substance each might be.

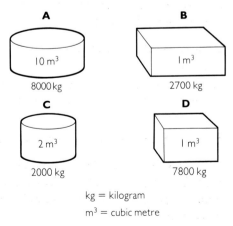

A
10 m³
8000 kg

B
1 m³
2700 kg

C
2 m³
2000 kg

D
1 m³
7800 kg

kg = kilogram
m³ = cubic metre

139

Pressure

Some forces are much more concentrated than others . . . . .

Pressure

Push a drawing pin into a piece of wood and the force on the wood is concentrated onto a small area. Push on the wood with your finger and the same force is spread over a much larger area.

The word *pressure* describes how concentrated a force is – the more concentrated a force, the higher the pressure.

Under the tip of the drawing pin the pressure is very high because the force acts on a very small area.

Under the finger the pressure is much lower because the force is spread over a much larger area.

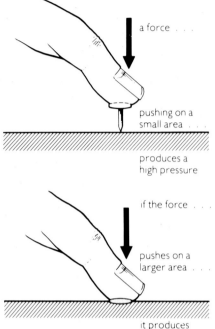

a force . . .

pushing on a
small area . . .

produces a
high pressure

if the force . . .

pushes on a
larger area . . .

it produces
a lower pressure

Calculating pressure   Pressure is measured in newtons per square metre, or *pascals* (Pa). It can be calculated using the equation:

$$\text{Pressure} = \frac{\text{force}}{\text{area}}$$

For example, if a 12 newton force pushes on an area of 2 square metres:

$$\text{Pressure} = \frac{12}{2} = 6 \text{ pascals}$$

Pressure in liquids

The force of gravity tries to pull a liquid downwards in its container. This causes pressure on the container and pressure on any object put into the liquid.

Pressure increases with depth   The deeper the liquid the greater the weight of liquid pushing on the bottom of the container. Dams are thicker at the base than at the top because they have to withstand a much higher pressure from the water at the bottom of a lake.

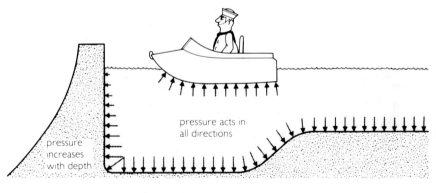

pressure acts in
all directions

pressure
increases
with depth

Pressure acts in all directions   A liquid under pressure pushes on every surface in contact with it no matter which way the surface is facing.

Pressure depends on the density of the liquid   Petrol is less dense than water. If petrol filled the lake instead of water the pressure at all points in the lake would be less.

**The pressure does not depend on the width of the
container** The pressure at the bottom of each glass due to the
weight of the liquid is exactly the same for each of them. The wider
mug does have a greater *weight* of beer to support but the downward
force from the beer is spread over a larger area.

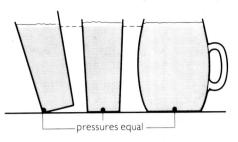
pressures equal

Hydraulics

Hydraulic machines make use of liquids under pressure rather than
levers and wheels. Liquids have two important properties which
make such machines possible:

1. Liquids cannot be squashed – they are virtually incompressible.
2. Press on a trapped liquid and the pressure is felt right through
the liquid.

Car brakes are operated hydraulically, so too are some car jacks.

The hydraulic jack The diagram below shows a simple working
model of a hydraulic jack:

The idea behind the jack looks
simple enough – if you push the
smaller piston down, the oil is
pushed through the pipe and
the larger piston is forced
upwards.

But the system has another
important feature. A small
downward force from your
thumb is all that is needed to
make the heavy brick move
upwards.

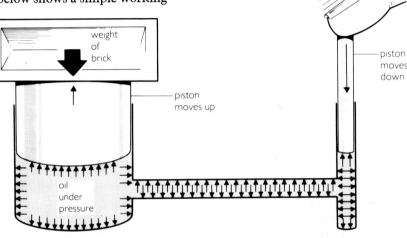

weight of brick

piston moves up

oil under pressure

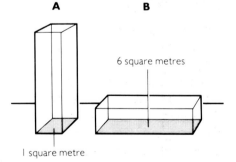

piston moves down

When the smaller piston is pushed down it puts the oil under high
pressure since the force is only being applied to a very small area.
High pressure oil pushing on the much larger area of the other
piston produces a large upward force – the force is strong enough to
overcome the weight of the brick so the brick is pushed upwards.

The upward *force* on the larger piston may be large but the upward
movement is not. The smaller piston has to be pushed down a long
way to lift the brick even a short distance.

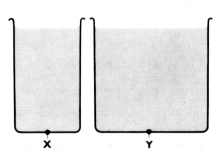

A   B

6 square metres

1 square metre

Questions

1. In the diagram on the right, both blocks **A** and **B** weigh 12
   newtons. Which block exerts most pressure on the ground?
   Why? Calculate the pressure under each block.
2. Why are dams thicker at the base than at the top?
3. The two cans on the right are filled with petrol. How does the
   pressure at X compare with the pressure at Y? If the petrol were
   replaced by water, how would this affect the pressures at X and Y?
4. What two properties of a liquid are used in a hydraulic jack?

X   Y

141

Floating and sinking

A steel boat floats in water, yet a lump of steel sinks. What makes things float? Why do some things float but not others?

Upthrust

A boat is kept afloat by an upward force from the water. The force is called an *upthrust* and it is caused by the pressure of the water pressing on the bottom of the boat.

Archimedes' principle The boat in the diagram takes up space previously filled with water – the boat has displaced water. Experiments show that:

the upthrust on the boat = the weight of water displaced

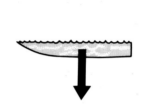

the upthrust on a boat is 9000 newtons   if the boat displaces water weighing 9000 newtons

This principle was discovered by Archimedes more than 2000 years ago. It applies to all objects in any liquid or gas, whether the objects are floating or not.

Floating

When a boat is first lowered into the water, the upthrust on it rises as more and more water is displaced. The boat stops sinking down into the water when the upthrust is just large enough to support the weight of the boat. When the boat is floating:

the upthrust on the boat = the weight of the boat

The law of floating If you study the forces in the diagrams above and on the right, you will see that:

the weight of the boat = the weight of water displaced

the upthrust is 9000 newtons

if the weight of the boat is 9000 newtons

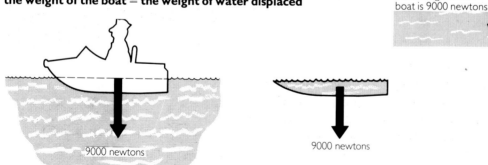

9000 newtons

9000 newtons

This principle applies to all objects floating in any liquid or gas and it is known as the *law of floating*.

Levels of floating

Two factors affect how low in the water a boat will float:

The weight of the boat and its load   If a boat carries a heavy load, more upthrust is needed to support the extra weight. The boat displaces more water by floating lower down in the water. If the load is too heavy, the upthrust cannot support the weight – the boat sinks.

The density of the water   Freshwater in a river is less dense than the salt water in the sea, and warm water is less dense than cold water – when water expands, its mass is more spread out.

This means that a boat must float lower in less dense water in order to displace the same weight of water. A boat floats lower in freshwater than it does in salt water, and a boat floats lower in warm water than it does in cold water.

To avoid the problem shown in the diagram, ships have a series of maximum loading levels ('Plimsoll' lines) marked on their sides to suit the different types of water in which they float.

The hydrometer

A hydrometer is a small float used for measuring the relative densities of liquids. Like a boat, a hydrometer floats lower in liquids of lesser density.

The stem of a hydrometer has a scale marked on it. The relative density of a liquid is found by reading the scale at the point where the liquid surface touches it.

The density of floating materials

You can tell whether a material will float or sink in a liquid by comparing its density with that of the liquid.

Ice is less dense than water – ice floats in water.
Warm water is less dense than cold water – warm water floats upwards through cold water.
Tar has the same density as water – tar just floats in water. Steel is more dense than water – steel sinks in water.
A steel ship floats in water because it is hollow – its average density is less than that of the water.

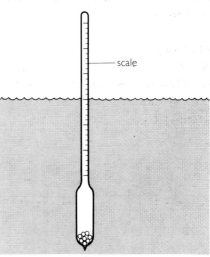
scale

Questions

1. What causes the upthrust on a boat?
2. A floating boat displaces water weighing 6000 newtons. What is the upthrust on the boat? What is the weight of the boat?
3. A girl climbs aboard a floating boat. What happens to the boat? What happens to the weight of water displaced? What happens to the upthrust on the boat?
4. Why does a boat float lower in freshwater than in salt water?
5. What is a hydrometer? What does it measure?
6. Steel sinks in water, yet a steel boat floats. Why?

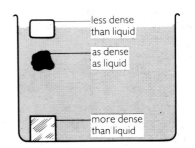

less dense than liquid

as dense as liquid

more dense than liquid

Pressure from the atmosphere

We live at the bottom of a deep ocean of air called the atmosphere. The air around us is at very high pressure – without this pressure you couldn't even make a decent cup of tea.

Atmospheric pressure

In some ways, the atmosphere is like a liquid – its pressure acts in all directions and its pressure gets less as you rise up through it. Unlike a liquid however, the atmosphere is much more dense at lower levels – the atmosphere stretches hundreds of kilometres into space yet most of the air lies within about 10 kilometres of the Earth's surface.

Atmospheric pressure at sea level   Down at sea level, the air pressure is about 100 000 pascals – that's equivalent to the weight of 10 cars pressing on each square metre.

We do not normally feel the effect of this crushing pressure. Like the undamaged oil can below, we have high pressure air inside us as well as outside.

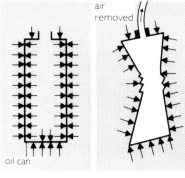

When the can is full of air, the air pressure inside the can balances the air pressure outside.

Remove the air in the can with a vacuum pump, and the sides of the can are pushed in by the pressure of the atmosphere.

Measuring atmospheric pressure

Instruments which measure atmospheric pressure are called *barometers*.

A simple barometer can be made by filling a long glass tube, sealed at one end, with a very dense liquid metal called mercury. The tube is put open end downwards into a dish of mercury and the level of the mercury in the tube starts to fall. The level stops falling when the pressure from the mercury left in the tube is balanced by the pressure of the atmosphere acting on the mercury in the dish. Atmospheric pressure is found by measuring the height of the mercury column on the scale – the result is usually expressed in 'millimetres of mercury'.

At sea level, atmospheric pressure is about 760 millimetres of mercury, though this varies slightly according to weather conditions. Rain clouds form in large areas of lower pressure air called depressions, so a fall in the barometer reading often means that bad weather is on the way.

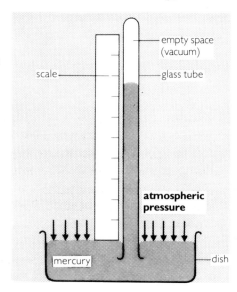

Using atmospheric pressure

The pressure of the atmosphere can be used in a variety of ways. Here are just two examples:

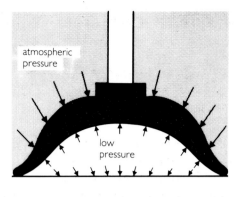

Rubber suckers   Push a rubber sucker against a piece of glass and you squeeze out most of the air from under the sucker. Try to pull the sucker away from the glass and the pressure of the trapped air drops as the air spreads to fill a larger space. With high pressure air outside it and low pressure air underneath it, the sucker is held firmly against the glass.

Drinking straws   When you 'suck' on a straw, you lower the air pressure in the straw by expanding your lungs.

Atmospheric pressure pressing down on the drink pushes the liquid up the straw and into your mouth.

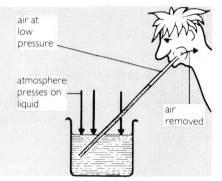

Atmospheric pressure and boiling

At sea level, water boils at about 100 °C. At the top of Mount Everest, water boils at only 70°C. So you can't even make a good cup of tea! The lower boiling temperature is a result of the lower atmospheric pressure at the top of the mountain.

If you watch water being heated, you can see small bubbles forming as the liquid starts changing into a gas. The pressure of the atmosphere on the water stops these bubbles growing in size, but the pressure in the bubbles grows stronger as the water gets hotter.

The water boils when the pressure in the gas bubbles is high enough to overcome atmospheric pressure. The gas bubbles grow rapidly and erupt on the surface of the water. If atmospheric pressure is low, the gas bubbles do not need to be so hot for this to happen – the water boils at a lower temperature.

Questions
1. Moving up through the atmosphere, what happens to:
   (a) the pressure of the air   (b) the density of the air?
2. Why does an old oil can not normally collapse under the pressure of the atmosphere? What would make the can collapse?
3. What instrument is used to measure atmospheric pressure?
4. Give the approximate pressure of the atmosphere at sea level in
   (a) pascals   (b) millimetres of mercury.
5. Why is it difficult to pull a rubber sucker away from a window?
6. What happens to the boiling temperature of water if atmospheric pressure falls?

Flight

Air pressure keeps aircraft and birds in the air. It keeps hot air balloons up as well, but in a rather different way.

Aircraft

When an aircraft is flying, there are 4 main forces on it:

Drag   is the force of air resistance. To reduce drag, the aircraft has a streamlined shape. The air flows smoothly over it.

Thrust   is the forward force from the engines. Thrust is needed to overcome drag.

Weight   is kept as low as possible by building the aircraft out of strong, light materials like aluminium.

Lift   is produced when the wings move through the air. The air speeds up as it flows over the top of the wings. This lowers its pressure. The pressure above the wings is less than the pressure underneath. So the wings are pushed up.

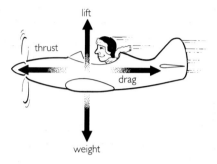

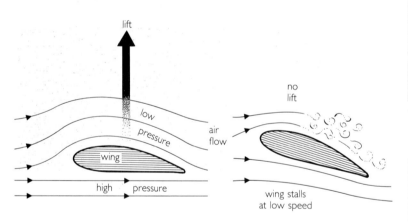

If a wing moves too slowly, the smooth flow of air over the top breaks up. The wing *stalls*. Lift is lost, and the aircraft starts to drop. The airliner in this photograph stalls at about 200 km per hour. A larger wing would give it more lift and a slower stalling speed. But the drag would be more.

Birds

Birds get lift from their wings. They get thrust as well. When they flap their wings, air is pushed backwards and downwards. This pushes the bird forwards. It also gives extra lift.

Like aircraft, birds need to be as light as possible. Hollow bones help to keep their weight down. So do wing feathers. These are strong and light, and give a large area for lift.

This 747 has enough lift to fly.

A large wing area gives this bird plenty of lift.

146

Floating in air

A hot-air balloon floats in air rather like a ship floats on water. There is an *upthrust* on it which supports its weight.

A balloon will float upwards if it weighs less than the air it displaces. That follows from the law of floating (page 142). The balloon in this photograph displaces over 2 tonnes of air. But when cold it has over 2 tonnes of air inside it. However, if the gas burner is lit, the air in the balloon heats up and expands. More than a ¼ of a tonne of air is pushed out of the hole at the bottom. The balloon becomes light enough to float.

Free fall

A skydiver jumps from her aircraft. Pulled by gravity, she accelerates towards the ground. As she gains speed, the air resistance on her increases and her acceleration gets less. When the air resistance balances her weight she stops accelerating. She won't speed up any more. She has reached her *terminal speed*.

Falling skydivers reach a terminal speed of about 60 m/s. It's slightly less if they spread their arms and legs to increase their air resistance. When they open their parachutes, their terminal speed drops to about 8 m/s.

Without air resistance, skydivers would all accelerate towards the ground at 10 m/s^2 – their speed would go up by 10 m/s every second. This is the *acceleration of free fall* near the Earth. It is the same for all falling things, light or heavy.

Almost enough upthrust to float.

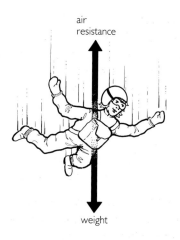

Sky-divers at terminal speed.

Questions

1. *thrust   drag   lift   weight*

   These are all forces on an aircraft. Which of them:
   (a) is caused by low pressure above the wings?
   (b) is the forward force from the engines?
   (c) is lost if the aircraft stalls?
   (d) is another name for air resistance?

2. Explain why the air in a hot air balloon has to be heated before the balloon will rise.

3. The skydiver in the diagram is falling at a steady speed. Say whether each of the following is *true* or *false*:
   (a) the skydiver is falling at her terminal speed
   (b) air resistance is less than the skydiver's weight
   (c) air resistance is the same as the skydiver's weight
   (d) without air resistance, a heavy skydiver would have the same acceleration as a light skydiver.

air resistance

weight

Further questions

1 *metre, newton, joule, kilogram, watt*
From the above list, name:
 a a unit of mass
 b a unit of force
 c a unit of energy
 d a unit of weight

2 An object is moved from the Earth to the Moon. How does this affect:
 a its mass
 b its weight?

3 In the following cases state whether the friction between the surfaces is an advantage or a nuisance:
 a friction between a wheel and its shaft
 b friction between brake pads and the disc on a car wheel
 c friction between shoes and the ground
 d friction between the pistons and the cylinders in a car engine
 e friction between car tyres and the ground.

4 As its fuel burns, a rocket motor is pushed forwards by a high force. What causes this force?

5 Why is a rocket motor able to work in space but a jet engine not?

6 Why is it an advantage for buses and lorries to have as low a centre of gravity as possible?

7 How much work would you do in lifting a stone weighing 20 N through a vertical distance of 2 m?
What type of energy does the stone gain as it is being lifted up?
What happens to this energy if the stone is dropped?

8 Copy out and complete the following:
 a In an electric kettle, electrical energy is changed into . . . . energy
 b In a . . . . electrical energy is changed into sound energy

9 Write down three ways, other than by direct heating, in which a liquid can be made to evaporate more rapidly.

10 Once the water in an electric kettle begins to boil, the temperature stops rising. If the kettle is left switched on for a minute or so, what happens to all energy supplied?

11 You are likely to chill quickly if you stand out in the wind in wet clothes. Why?

12 Why do you feel more uncomfortable on a humid day than on a dry day?

13 How would you measure the volume of a rubber bung?

14 Petrol has a density of 800 kg/m³. What is the mass of 3 cubic metres of petrol?

15 **a** What volume of air is there in a room 5 metres long, 5 metres wide, and 2 metres high?
 b If the density of air is 1.5 kg/m³, what is the mass of the air in the room?

16 When you press a drawing pin into wood, why is the pressure on the wood very much greater than the pressure from your thumb?

17 **a** What is the instrument in the diagram called?
 b If 1 millimetre in the diagram represents 1 centimetre on the actual apparatus, what is the pressure of the atmosphere in centimetres of mercury?
 c In what way would the reading change if the instrument were taken up a mountain?

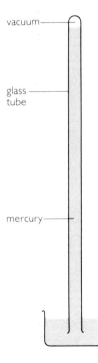

vacuum

glass tube

mercury

18 The pressure inside a pressure cooker is higher than the air pressure outside. How does this affect the boiling point of the water in the pressure cooker?

19 Describe 2 ways in which the pressure of the atmosphere can be put to practical use.

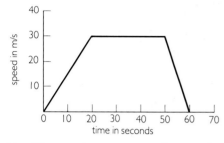

20 The graph shows how the speed of a motorcycle changed with time over a certain journey.

a For how long did the motorcycle accelerate?
b What was the maximum speed of the motorcycle?
c For how long did it travel at this maximum speed?
d What was the deceleration of the motorcycle just before it stopped?

21 The diagram shows a simple bottle opener being used to remove the top from a bottle.

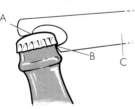

a The bottle opener is being used as a lever. Which of the points A, B, or C, is the fulcrum (turning point) of the lever?
b At which point A, B, or C, is the *load* applied to this lever?
c Is the force on the bottle top at point B *greater* or *less* than the force used at C? Give a reason.
d How could you improve the design of the bottle opener so that less force is required to remove the bottle top?   SEG/SWEB

22 Using the information given in the table below, answer the questions which follow.

Substance	Melting point in °C	Boiling point in °C	Density in kg/m³
A	− 38.9	356.7	13 546
B	− 7.1	58.9	3120
C	197.5	877.0	970

a Which substance would:
i float in water (density 1000 kg/m³)?
ii be liquid at normal room temperature?
iii be most readily turned into vapour?
b Calculate the volume of 6240 kg of substance B.   SEG

23

Jane has a motorcycle with a 125 cm³ engine.
a 125 cm³ is often called the capacity of the engine.
i What word in Science would you use instead of capacity?
ii Jane's motor cycle is on a work bench. She is given some oil and a measuring cylinder. How could she check the 'capacity' of the engine?
b The engine is made of aluminium. One reason that aluminium is used is because of its low density.
i What is meant by the density of a material?
ii Describe how you would measure the density of aluminium using the equipment available in your laboratory. **iii** Why is it important to build a motorcycle with low density materials? **iv** Suggest one *other* advantage of an aluminium engine.   SEG/SWEB

24 The chart shows the energy content of Tracey's evening meal.

Food	Energy in 100 g (in kJ)	Mass of 1 portion (in g)
sausages	1300	100
chips	1000	150
peas	300	100

a What do 'g' and 'kJ' stand for?
b Which of the three foods has most energy in 100 g?
c Which of the three foods has most energy in a portion?
d What is the total energy of Tracey's meal?

25 This question is about a car fitted with a four-stroke petrol engine.
a Copy and complete the table below to show what is happening to the valves and pistons in one cylinder of the engine.

Stroke	Inlet valve **open** or **shut**	Outlet valve **open** or **shut**	Piston moving . . . **up** or **down**
induction			
compression			
power			
exhaust			

b What is taken into the engine during the induction stroke?
c When does the plug 'spark'?
d Which of the following describes the energy changes taking place in the engine?
    A heat → chemical → kinetic
    B potential → kinetic → heat
    C chemical → heat → kinetic
    D chemical → kinetic → potential
e The car could have been fitted with a diesel engine instead of a petrol engine. What are the advantages of a diesel engine? What are the disadvantages?

26 Debbie and Joe found a long plank of wood which they balanced across a log. The diagram shows Debbie and Joe balanced on the plank. (The weight of 1 kg is 10 N.)

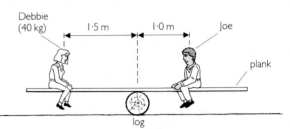

a If Debbie has a mass of 40 kg, how much force does she exert on the plank?
b If she sits 1.5 m from the pivot (turning point) what is the moment of this force about the pivot?
c Joe balances Debbie when he sits only 1.0 m from the pivot. What is Joe's weight?
d What is Joe's mass?   SEG/SWEB

27 This question is about a hydroelectric power scheme. Water rushing down from the lake turns turbines at the base of the dam. The turbines turn electrical generators.

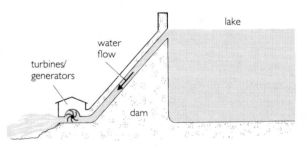

a Why is the dam thicker at the base than at the top?
b Where does the water have most potential energy?
c Where does the water have most kinetic energy?
d The hydroelectric scheme is a renewable energy source. What does 'renewable energy source' mean?
e Hydroelectric schemes don't pollute the air. But they can cause other problems. Explain.

Chapter 5

Living things

What's the difference between animals and plants?
What are they both made from?

Animals and plants

Animals and plants are living organisms. They are alike in many ways but there are differences in the substances they contain and the ways in which they get their food. Food supplies animals and plants with the energy they need to live.

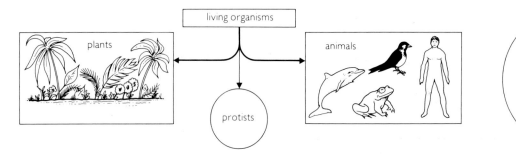

protists are simple organisms which are difficult to class as either animals or plants

Similarities between animals and plants
The following are features of all living things:

Feeding   Animals take in food. Plants take in materials to make their food.

Respiration   Animals and plants need energy to move, to grow and to maintain life. Usually, they get the energy they need by combining their food with oxygen.

Excretion   Animals and plants produce waste materials which they must get rid of in some way. Human beings do so in breathing out, sweating, and using the lavatory.

Growth   Animals and plants may grow bigger. They may also grow more complicated in their structure.

Movement   Animals and plants are able to move, though animals usually make more obvious movements than plants.

Reproduction   Animals and plants can produce others of their own kind. Human beings have children, a simple organism may just split into two. Life goes on even though individuals die.

Sensitivity   Living things react to the outside world. A plant may search for water or light, an animal may seek warmth. Animals usually respond to a stimulus much more quickly than plants.

Differences between animals and plants
Animals and plants differ in many ways. These differences include:

Cellulose   Plants are partly built from a substance called cellulose. Animals do not use cellulose for body-building.

Chlorophyll   Most plants contain a green substance called chlorophyll. Plants use chlorophyll to absorb energy from sunlight.

Eating sunshine, or eating plants?

Green plants make their own food from carbon dioxide, water and other simple materials. They can only do this by using energy from sunlight to turn these substances into new plant tissue and food for storage. The process is called *photosynthesis* and oxygen is 'breathed out' from plants as it takes place.

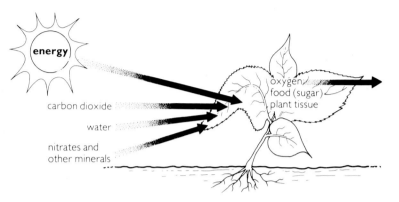

Animals cannot obtain energy directly from sunlight. They take in food which has been made and stored in plants. Some animals feed directly on green plants, others feed on animals which eat green plants.

Energy for living

Like plants, animals get the energy they need by combining their food with oxygen during a process called *respiration*. They breathe in oxygen so that respiration can take place in the tiny cells which make up their bodies. As it takes place carbon dioxide and water are formed.

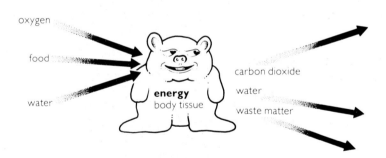

How many 'signs of life' can you spot in this quiet rural scene?

Questions

1. What are the seven features common to all living things?
2. What substance makes plants green? How do plants use it? What other substance is found in plants but not animals?
3. What substances do plants take in to make their food? Where does the energy come from to make the food? What is the process called? What gas does it produce?
4. What is respiration? Where does it take place?
5. Energy is released during respiration. What else is produced?
6. What gases do you breathe in and out?

Living cells

All animals and plants are made up of tiny living units called cells. These cells come in many different shapes, sizes and types. Within them all the basic chemistry of living and growing takes place.

Cell structure: animals

Most cells are so small that many thousands would fit on the head of a pin. Not all cells are so small – a chicken's egg is one single cell.

A thin skin called a *membrane* surrounds each cell. Within it are the jelly-like *cytoplasm* and *nucleus*.

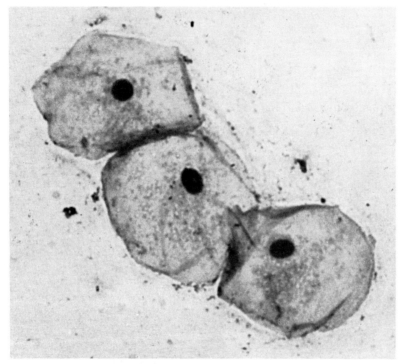

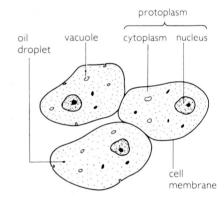

Typical animal cells. The photograph shows cells from the lining of the cheek.

Cell membrane   This thin skin controls the flow of all the substances which pass in and out of the cell.

Cytoplasm   This is the chemical factory of the cell. Here, new substances are built up from materials taken into the cell and energy is released and stored. Sometimes the cytoplasm contains tiny droplets of liquid – these are known as *vacuoles*.

Nucleus   The nucleus controls all the chemical reactions that take place in the cell. Thread-like *chromosomes* in the nucleus store the chemical instructions used to build the cell.

All the living matter in a cell is known as *protoplasm*.

Cell structure: plants

Plant cells also have a *cell membrane*, *cytoplasm* and *nucleus* but they differ from animal cells in several ways.

Cell wall   Plant cells are surrounded by a firm wall of *cellulose*. Cell walls hold plant cells together and give plants much of their strength.

Chloroplasts   Many plant cells contain tiny structures called *chloroplasts*. Within them is the green substance called *chlorophyll* which absorbs energy from sunlight.

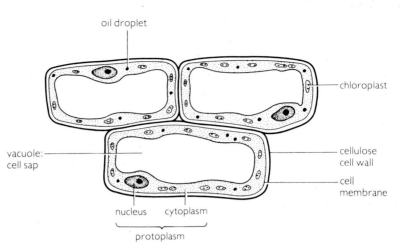

Cell-sap   Plants cells contain a large central droplet (a *vacuole*) of watery liquid called *cell-sap*. Osmotic pressure from this liquid pushes on the outer parts of the cell keeping the cell firm or *turgid*.

Cell specialisation

Advanced organisms (like you and me) are built up from many billions of cells. Different groups of cells have different jobs to do – the cells are *specialised*. For example, one group of cells form muscle to move the body; another group will be concerned with reproduction.

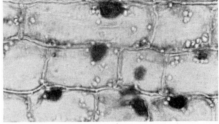
Plant cells.

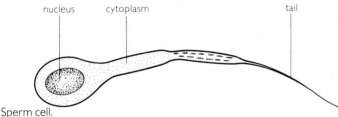
Sperm cell.

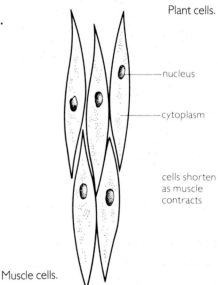
Muscle cells.

Groups of similar cells form *tissue*. Tissues grouped together to do one particular job form an *organ*. Muscles are organs, so are eyes and lungs.

Questions

1. What controls the flow of substances in and out of a cell?
2. Which part of a cell controls all chemical reactions taking place in the cell?
3. Where do the main chemical reactions in a cell take place?
4. How is the outside of a plant cell different from that of an animal cell?
5. What do chloroplasts contain? Why?
6. What do the large liquid droplets in a plant cell contain? What are they called?
7. What is tissue? What is the name given to a group of tissues which do one particular job?

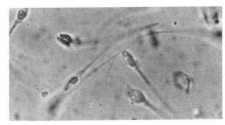
Human sperm cells, enlarged 1500 times.

155

Cells and more cells

The amoeba is a tiny single-celled creature which reproduces by dividing to form two new cells. Advanced plants and animals reproduce by joining together a male and female cell to form a single new cell. By dividing over and over again, in different ways, the single cell becomes the many different types of cell needed to make up the new plant or animal. This incredible dividing process is controlled by chromosomes in the nucleus of each cell.

The amoeba

The single-celled amoeba lives in ponds and damp soil and feeds on microscopic plants. It grows bulges around small pieces of food which it then takes into its cytoplasm in the form of small liquid droplets (vacuoles). Digested food is absorbed into the cytoplasm. Undigested waste is left behind as the amoeba 'flows' on its way.

Binary fission   The amoeba reproduces itself by a process called *binary fission*. The nucleus of the cell divides in two and the cytoplasm then separates to form two new cells. These *daughter* cells are copies of the original *parent* cell.

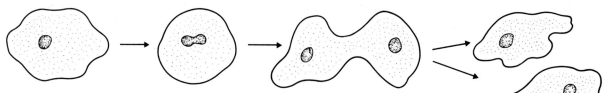

parent cell   nucleus divides . . .   . . . then cytoplasm

daughter cells formed

Reproduction in cells

Asexual reproduction   Like the amoeba, many organisms reproduce by dividing to form new organisms. Reproduction of this kind is described as *asexual*.

A tulip reproduces asexually when new bulbs grow on the side of the original bulb and then separate from it.

Sexual reproduction   A new human life starts when a *sperm cell* from a man combines with an *egg cell* in a woman to form a single new cell called a *zygote*. This new cell divides over and over again to form the billions of cells in a baby's body. The body has many different parts – every 'daughter' cell does not end up an exact copy of its parent cell.

This process is called *sexual reproduction* and it is used by the more advanced plants and animals. Many common plants can reproduce both sexually and asexually.

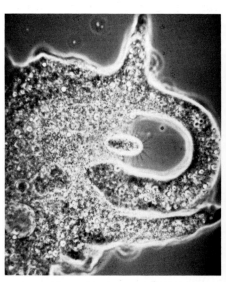

This amoeba has just finished bulging round its prey, taking in some liquid with it.

Chromosomes: building new cells

Cells are made up of molecules, some of which contain many thousands of atoms. Cells build new molecules by rearranging the atoms of incoming material.

The chemical building instructions are stored in each nucleus in coded form by molecules of deoxyribonucleic acid (*DNA*). These DNA molecules can produce exact copies of themselves, so their building instructions can be passed on to any new cells that form.

DNA is stored in the nucleus of a cell in structures called *chromosomes*. Our own cells each contain 46 chromosomes which between them store the chemical instructions for building the complete human body.

Mitosis   Before a cell divides, copies of each chromosome form alongside the originals and then separate from them. As the cell divides, a full set of chromosomes collects at each end to form part of two new nuclei. The process is called *mitosis:* it ensures that full chemical building instructions are passed on to the new cells.

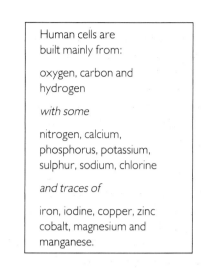

Human cells are built mainly from:

oxygen, carbon and hydrogen

with some

nitrogen, calcium, phosphorus, potassium, sulphur, sodium, chlorine

and traces of

iron, iodine, copper, zinc cobalt, magnesium and manganese.

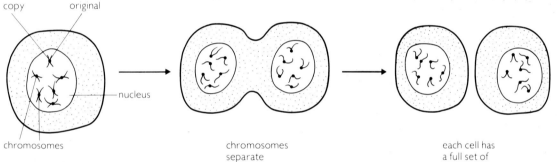

Meiosis   The cells that go to form reproductive tissue – sperm in males, and eggs in females – split so as to have only 23 chromosomes each. This type of splitting is called *meiosis.* In the original cell from which you grew, called the *zygote,* 23 chromosomes come from your father, and 23 from your mother. So you inherited features from both parents.

Questions

1. How does an amoeba take in and then store its food?
2. The cytoplasm of an amoeba is divided into two regions: what are they called?
3. How does an amoeba reproduce itself? What is the process called?
4. What three types of atom are human cells mainly built from?
5. What structures in the nucleus of a cell store its chemical building instructions? What happens to these structures before a cell divides? What substance carries the instructions in coded form?

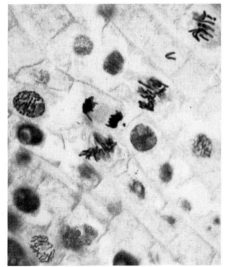

This photo is of the very tip of an onion root. The cells are dividing by mitosis.

Green plants: making and using food

Like other living things, green plants need food in order to live and grow. They take in simple materials from the soil and the atmosphere and turn these into food by using the energy in sunlight.

Transpiration and photosynthesis

Transpiration   Large amounts of water evaporate from the leaves of a plant, though the leaf cells do have some control over the rate at which water is lost. This loss of water by evaporation is known as *transpiration* and it draws new water and vital minerals from the soil, up through the plant, to the leaves.

The upward flow of water through a plant is called a *transpiration stream*. In a large tree the flow of water may be more than 100 litres every day.

Although most of this water evaporates, some is kept in the plant and used in the food-making process called *photosynthesis*.

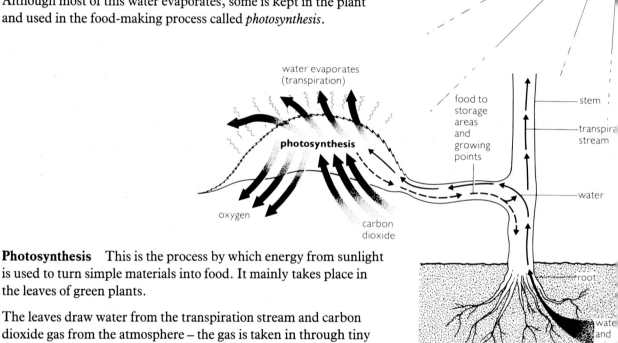

Photosynthesis   This is the process by which energy from sunlight is used to turn simple materials into food. It mainly takes place in the leaves of green plants.

The leaves draw water from the transpiration stream and carbon dioxide gas from the atmosphere – the gas is taken in through tiny pores called *stomata*.

Chlorophyll in the leaves absorbs energy from sunlight and cells use this energy to turn the carbon dioxide and water into foods such as sugar:

carbon dioxide + water + energy → food (sugar) + oxygen

Oxygen released by this chemical reaction passes out into the atmosphere through the tiny pores in the leaves.

One of the simplest foods made by plants is a type of sugar called *glucose*. Chains of glucose molecules form the tough substance called *cellulose* of which plants are largely made.

Stored food   Much of the food made in a leaf is carried away in solution to other parts of the plant. Some goes to points of new growth where it is used in building the plant – some goes to storage areas in the roots, stem or leaves. Stored food can later be used as a source of energy.

Energy from food: respiration

Plants need energy to drive all the chemical reactions which enable them to live and grow. *Respiration* is the process by which plants release energy from their food:

food + oxygen ⟶ carbon dioxide + water + energy

Respiration takes place all the time, so plants need a steady supply of oxygen.

During daylight hours, plants produce more oxygen by photosynthesis than they can use. They therefore give out oxygen into the atmosphere.

Photosynthesis stops at night – plants must then draw the oxygen they need for respiration from the atmosphere.

Overall, plants produce more oxygen than they can use. Animals use up this oxygen, so the amount of oxygen in the atmosphere remains steady.

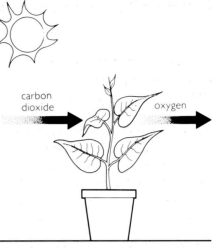

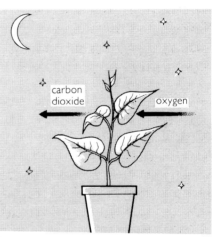

Testing for starch

During the daytime, photosynthesis may produce sugar at a faster rate than it can be used or carried away from the leaf. Many plants convert this extra sugar into *starch* which is temporarily stored in their leaves. The diagram shows a simple test you can perform on a leaf to find out whether or not starch is present:

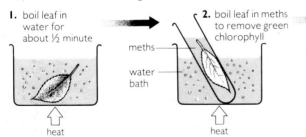

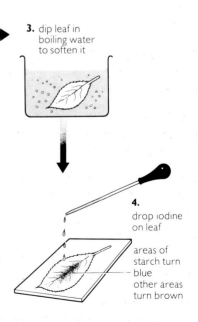

Questions
1. By what process is water drawn up through a green plant?
2. Where in a plant does food-making mainly take place? Why does it only take place during daylight hours?
3. What happens to the food made in a plant?
4. Why do plants need energy?
5. What gas do plants need to release the energy from their food?
6. When does a green plant give out oxygen?
7. When does a green plant take in oxygen?
8. Why does the oxygen in the atmosphere remain steady?
9. Why may starch build up in a leaf during the daytime?

159

Flowering plants: structure and growth

Plants have a structure which enables them to work as efficient food makers. The food supplies the plant with energy – it also supplies it with the materials it needs in order to grow.

Structure of a flowering plant

A typical flowering plant has a *root* in the ground and a stem which supports leaves, buds and flowers. The upward growing stem with its leaves, buds and flowers is known as a *shoot*.

Leaf   Leaves are the 'food factories' of a plant. They are supplied with simple materials from the atmosphere and the soil. They use the energy in sunlight to turn these materials into food.

The upper surface of a leaf is often covered with a waxy layer called *cuticle* which stops the leaf losing too much water by evaporation. Just under this layer are the food-making cells which absorb energy from sunlight.

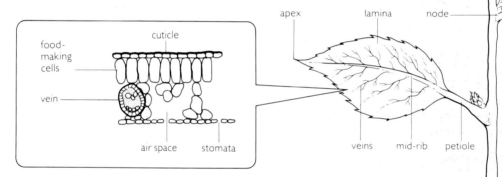

The food-making cells get their supply of carbon dioxide through tiny pores (*stomata*) which are mainly on the underside of the leaf. Both oxygen and evaporating water leave the leaf through the same pores.

A system of narrow tubes brings water and minerals to the food-making cells and takes dissolved food away. The tubes are bunched together to form *veins* – each leaf contains a network of veins which meet to form a *main vein* through the centre of the leaf.

Stem   The stem supports and spaces out the leaves so that they can receive as much light as possible. Food-tubes and water-tubes run through the stem. They connect the food- and water-tubes in the leaves with those in the roots.

Roots   Roots fix a plant firmly in the soil. They take in water and minerals from the soil through tiny tubes called *root hairs*. In many plants, the roots act as a store for much of the food made in the leaves.

160

New growth

Some of the materials made and stored in a plant are carried through the food-tubes to points of new growth. Here, new plant tissue is built up as rapid cell division takes place.

Growth in roots   There is a *growing point* at the tip of each root where new root tissue is rapidly built up. As the root tip grows, the root pushes itself further into the soil in search of water.

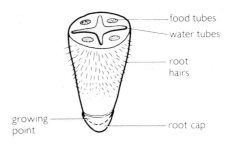

A plant root.

Growth in buds   In a bud, young, folded leaves are packed around a short stem and protected by tough outer leaves called *bud scales*. At the end of the short stem there is a flower, or a growing point where a new bud will form.

The bud at the top of the main shoot is called the *terminal bud*. In the springtime, the leaves in this bud unfold and space out as the stem pushes upwards. If the terminal bud produces a flower, no growing point is left when the flower finally falls away. Growth in the following season then takes place in the side or *axillary* buds.

A plant bud.

Thickening in the stem   As a plant grows its stem thickens in order to support the extra weight above it.

New cells are made in a thin layer called the *cambium* which lies between the food-tubes and the water-tubes in the stem. The cambium moves outwards as new food-tubes are formed on the outside of the layer and new water-tubes on the inside.

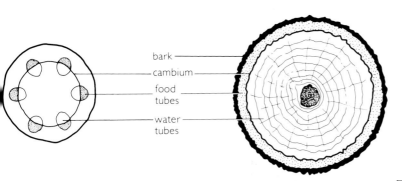

Questions

1. Why do some leaves have a waxy layer on their upper surface? What is this layer called?
2. Where in a leaf are the food-making cells?
3. How does carbon dioxide get into a leaf?
4. What are veins? Why does a leaf contain them?
5. Give two reasons why a plant needs (a) a stem (b) roots.
6. Where in a root is the growing point?
7. What does a bud contain?
8. Why does the stem of a plant thicken as the plant grows?
9. Stems grow thicker as a result of cell division. Where does this take place?

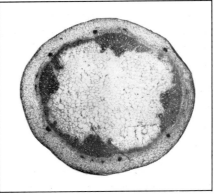

Cross section through the stem of a sunflower.

Reproduction in flowering plants (1)

Plants which flower do so in order to reproduce themselves. Flowers contain the tiny male and female cells which, when combined, grow into seeds. These seeds are scattered over a wide area. Some survive to grow into new plants.

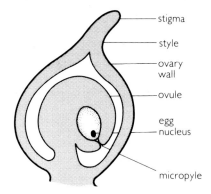

Carpel: before fertilisation

Flower structure

Some flowers contain male cells, some contain female cells but many contain cells of both types. The male and female cells are stored in the *stamens* and *carpels* which stand in rings around the centre of a flower.

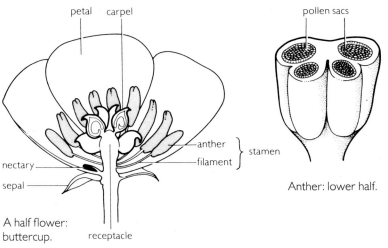

A half flower: buttercup.

Anther: lower half.

Stamens   These hold the grains of *pollen* which contain the male cells. The pollen is stored in *pollen sacs* in the bulge at the end of each stamen. Pollen is exposed when this bulge (the *anther*) ripens and splits.

Carpels   Each carpel has a hollow in its base called an *ovary*. From the walls of this hollow grow tiny round *ovules* which each contain one female cell – an *egg-cell*. When the tip of a carpel is ripe, it is ready to receive and nourish grains of pollen.

Pollination and fertilisation

For a male cell to combine with a female egg-cell, pollen grains must first be transferred to the tip of a carpel. This process is called *pollination*.

Cross-pollination   This is the transfer of pollen grains from one plant to carpels on another plant of the same type. In some cases the pollen grains are carried by wind, in others by insects such as bees which move from flower to flower. Flowers are often brightly coloured to attract insects. The insects visit them in search of the sugary nectar in the nectaries and pick up grains of pollen on their bodies as they do so.

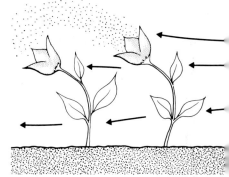

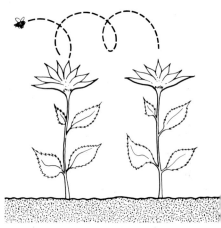

Pollination by wind and insects.

Self-pollination This is the transfer of pollen grains from stamens to carpels in the same plant.

Most plants arrange things so that self-pollination is not very likely to happen. Self-pollinated plants have much less chance of producing variations that will help future plant generations in their struggle for survival. Pollen sacs and carpel tips may ripen at different times to prevent self-pollination taking place.

Fertilisation When ripe, the tip of a carpel will nourish grains of pollen which stick to it. Each pollen grain then grows a long *pollen tube* which pushes down towards an ovule. A nucleus from a male cell passes down this tube and combines with the nucleus in the female egg-cell – the egg-cell is then *fertilised*. After fertilisation of the egg-cells has taken place, much of the outer flower shrivels up and falls off.

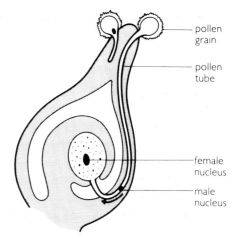

Carpel: during fertilisation.

Seeds and fruits

Seeds Within its ovule each fertilised egg-cell grows by cell division to form a seed. Each seed is a tiny new plant with a thick protective *seed coat* around it.

Fruits Biologists describe the complete ovary after fertilization as a fruit – they do not use the term fruit in quite the same way as a greengrocer. Oranges are fruits but so too are tomatoes and bean pods.

When you eat an apple you are actually eating the *receptacle* which contains the ovary. This receptacle is often called a *false fruit*.

Dispersal of seeds and fruits Plants scatter their seeds over a large area to give as many of them as possible a chance of surviving as new plants. Different plants use different methods to disperse their seeds.

Some seeds are carried by wind, some hook on to animals. Some seeds come from pods which pop and flick their contents across the ground. Some seeds remain in fruits which are then eaten by animals – the hard seeds pass out onto the ground with the faeces.

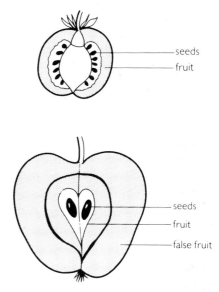

Two fruits: tomato and apple.

Questions

1. Where in a flower are the male cells of reproduction stored? Where are the female egg-cells stored?
2. What is the tip of a carpel called? What name is given to the bulge at the end of a stamen?
3. Give two ways in which cross-pollination can take place.
4. Why are some flowers brightly coloured?
5. What is self-pollination?
6. A pollen grain attaches itself to the tip of a carpel. How does fertilisation then take place?
7. What is a seed? What is a fruit?
8. List the ways in which fruits and seeds can be dispersed.

Reproduction in flowering plants (2)

Given the right conditions, a seed will grow into a new plant. In time, this plant will build up the store of food it needs in order to reproduce itself. Some plants do not survive after they have reproduced but many types live on to reproduce over and over again.

Germination

Seeds are usually very dry when scattered from a plant. In this state some can remain alive but inactive for long periods – they are then *dormant*.

Growth begins when a seed is in the presence of water, oxygen and a temperature that suits it. When a seed begins to grow, it is said to *germinate*.

The germination of a broad bean seed is described below though many other seeds germinate in a similar way.

Germination of a broad bean seed   Inside the seed coat of a broad bean seed lies the beginnings (the *embryo*) of a new plant. A tiny leafy shoot is present, so too is a tiny root. These are joined to two *seed-leaves* which store the food needed for the first stages of growth.

Germination begins when the bean seed absorbs water, swells, and bursts through the seed coat:

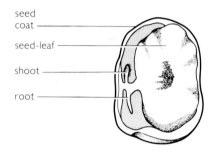

A broad bean seed.

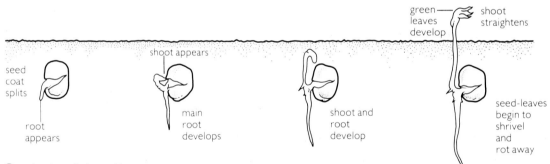

Germination of a broad bean.

Tropisms

No matter how a seed falls into the soil, its shoot grows upwards and its root grows downwards – shoot and root seem to sense the direction of gravitional pull as they grow.

Growth in a direction controlled by gravity or some other influence is called a *tropism*. Light also affects the direction of growth – young plants placed indoors but near a window will grow with their stems bending towards the light.

Life cycles

As it grows, a flowering plant passes through two distinct phases.

During its *vegetative phase* the plant gains in size and weight as it builds up a supply of stored food. During its *reproductive phase* the plant uses this stored food to produce flowers, seeds and fruits.

Some plants pass through each phase only once before they die. Some live through many cycles of growth and reproduction.

Annual plants   (e.g. marigolds) These live through one growing season only. They reproduce once in the year and then die as their food reserves are exhausted.

Biennial plants   (e.g. cabbages, carrots) These spend the growing season in their first year building up a store of food in their roots, stems or leaves. These food stores survive the winter, unless you eat them, and the food is used to produce flowers, seeds and fruits the following year. The plant then dies.

Herbaceous perennials   (e.g. crocuses) These die away above ground level each autumn but build up enough stored food to flower every year or so. The food is stored in swollen stems, roots or leaves known as *bulbs, corms, rhizomes* or *tubers.*

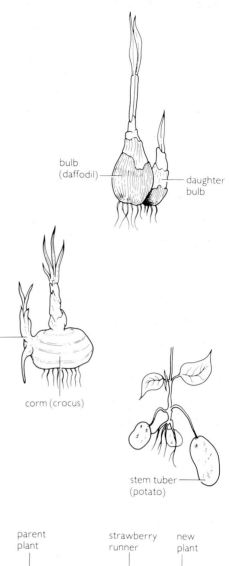

bulb (daffodil)

daughter bulb

daughter corm

corm (crocus)

rhizome (iris)

stem tuber (potato)

Woody perennials   (e.g. trees, shrubs and bushes) These have woody stems which last from year to year.

Vegetative reproduction

Although many plants form flowers in order to reproduce, some perennial plants are also able to reproduce by growing parts which separate from the parent plant to form new plants. Side buds in a daffodil bulb grow into bulbs which may separate from the parent bulb. Strawberry plants produce runners which take root to form new plants.

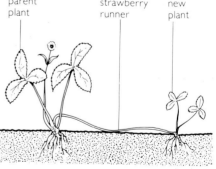

parent plant

strawberry runner

new plant

Reproduction of this type is known as *vegetative reproduction* – it is a form of asexual reproduction.

Questions

1. What does a seed need in order to germinate?
2. Describe the stages in the germination of a broad bean.
3. What are tropisms? Give two examples.
4. What two phases are there in the life cycle of a flowering plant? What happens during each?
5. Where do herbaceous perennials store their food?
6. How can a strawberry plant reproduce itself asexually?

Classifying living things

Scientists begin to classify living things by grouping them into *kingdoms*.

Most living things can be placed without too much difficulty in either the *plant kingdom* or the *animal kingdom*. Some do not fit easily into either – many scientists place these in a third kingdom, the *protista*.

The three kingdoms are each divided into smaller groups of organisms called *phyla*, some of which are shown in the charts below. The organisms in any one *phylum* will have several general features in common.

The smallest groups of living things are called *species*. Members of a species are so much alike that they can mate together and produce young like themselves. Humans are a species, so are barn owls, common meadow buttercups and guinea pigs. There are well over a million species of living organism on Earth.

The protista kingdom

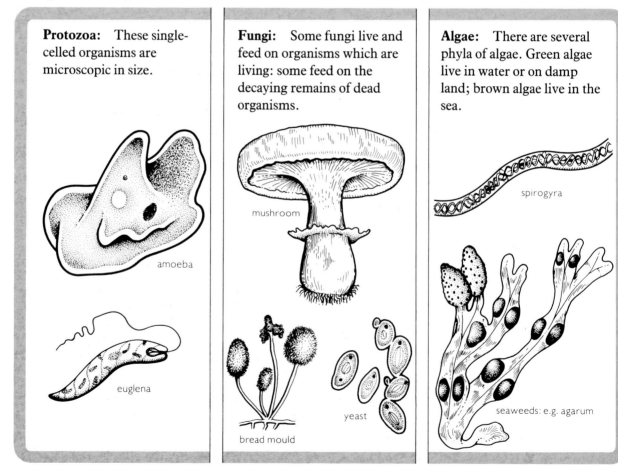

Protozoa: These single-celled organisms are microscopic in size.

amoeba

euglena

Fungi: Some fungi live and feed on organisms which are living: some feed on the decaying remains of dead organisms.

mushroom

bread mould

yeast

Algae: There are several phyla of algae. Green algae live in water or on damp land; brown algae live in the sea.

spirogyra

seaweeds: e.g. agarum

The plant kingdom

Bryophytes: These plants live in damp places and have no true roots, stems or leaves though they possess structures which look rather similar.

The bryophytes include the **liverworts** and the **mosses**.

Ferns: Ferns have roots, stems and leaves but they produce no flowers.

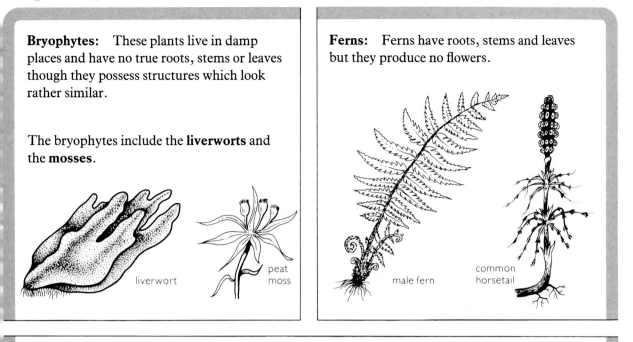

liverwort

peat moss

male fern

common horsetail

Seed plants: The seed plants are the most advanced forms of plant life to be found on Earth.

Flowering seed plants: These produce their seeds in ovaries which then form fruits.

Monocotyledons: are narrow-leaved plants which contain only one seed-leaf in each seed.

Dicotyledons: have broad leaves, and seeds which each contain two seed-leaves. They include:-

herbaceous plants e.g. buttercups and daisies;
shrubs e.g. woody hedges and bushes;
deciduous trees e.g. oak, ash and elm.

Non-flowering seed plants: These include coniferous trees like the fir, pine and spruce.

The seeds are contained in cones rather than in fruits.

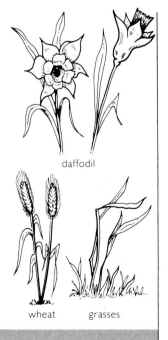

daffodil

wheat   grasses

buttercup   oak

spruce

Classifying living things: the animal kingdom

Coelenterates:
Nearly all live in the sea.

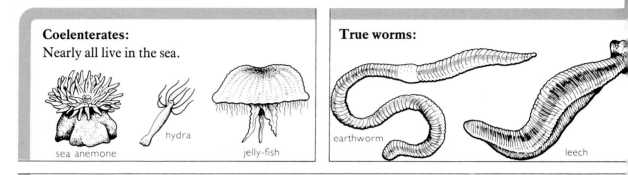

sea anemone — hydra — jelly-fish

True worms:

earthworm — leech

Arthropods: This is the largest phylum in the animal kingdom. Arthropod bodies are in sections, with a strong outer case called an *exoskeleton*. Arthropod legs are in pairs and jointed.

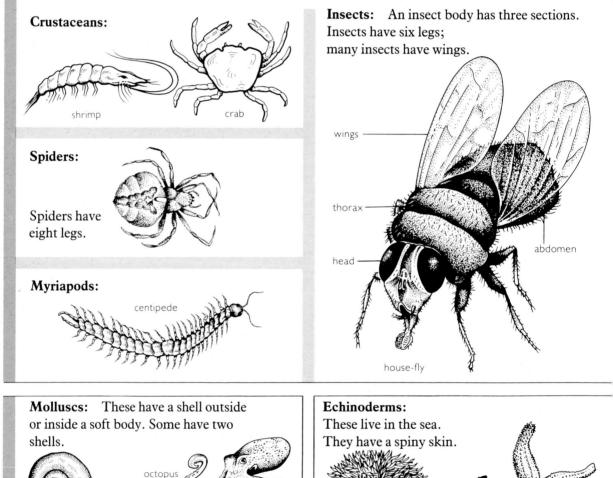

Crustaceans:

shrimp — crab

Spiders:

Spiders have eight legs.

Myriapods:

centipede

Insects: An insect body has three sections. Insects have six legs; many insects have wings.

wings
thorax
head
abdomen

house-fly

Molluscs: These have a shell outside or inside a soft body. Some have two shells.

land snail — octopus

Echinoderms:
These live in the sea.
They have a spiny skin.

sea urchin — starfish

Vertebrates:

These animals have a backbone (a vertebral column).

Some animals have a body temperature which changes – 'cold blooded' animals:

Fish: Some fish have a cartilage skeleton.

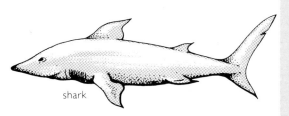
shark

Some fish have a skeleton made of bone.

perch

Amphibians: These live on land. They lay eggs, and produce young which live in the water at first.

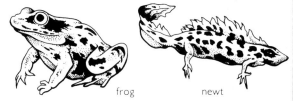
frog   newt

Reptiles: These have a scale-covered skin. They are egg-laying.

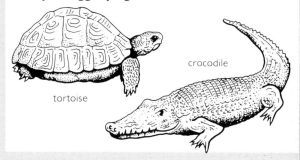
tortoise   crocodile

Some animals have a constant, warm, body temperature – 'warm blooded' animals:

Birds: Birds are feather covered. Their front limbs form wings. Most, but not all, birds can fly.

Mammals: Mammals are hair covered. They suckle their young with milk from the mother. Primates (monkeys, apes, humans) are the most advanced animals of all.

robin   penguin   dolphin   mouse   human

Herbivores: animals which feed on plants, e.g. cows, horses.
Carnivores: animals which feed on the flesh of other animals, e.g. lions, owls.
Omnivores: animals which feed on plants and animal flesh, e.g. humans.

Evolution

The first living organisms appeared on Earth more than 3000 million years ago. As generation followed generation these organisms slowly changed and in time they developed into the animals and plants of today. This process of gradual development is called evolution – an explanation of the process of evolution is offered by the theory of natural selection.

Evolution of the mammals

The diagram shows some of the stages in the evolution of the mammals which exist today.

The path of evolution has divided many times in the past. Where animals have begun to live and breed in separate groups, evolution has taken a different course with each group.

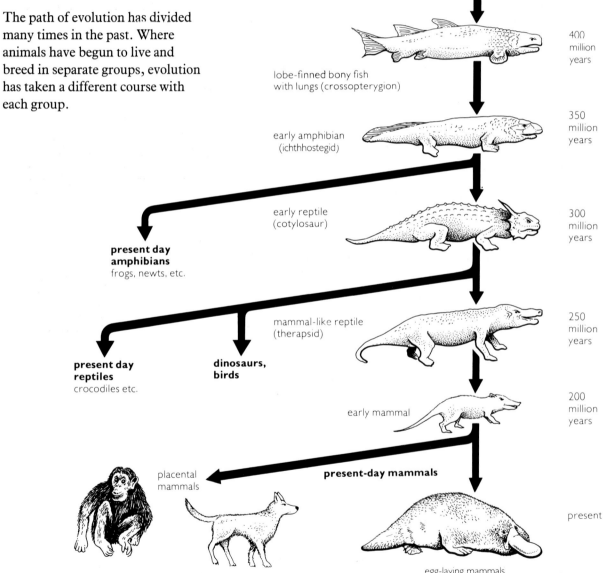

lobe-finned bony fish
with lungs (crossopterygion)

400 million years

early amphibian
(ichthhostegid)

350 million years

early reptile
(cotylosaur)

300 million years

present day amphibians
frogs, newts, etc.

mammal-like reptile
(therapsid)

250 million years

present day reptiles
crocodiles etc.

dinosaurs, birds

early mammal

200 million years

placental mammals

present-day mammals

present

egg-laying mammals
(e.g. duck-billed platypus)

The theory of natural selection

Charles Darwin suggested that evolution was the result of a process which he called *natural selection*. The *theory of natural selection* is explained in stages below – for simplicity, the explanations deal only with animal life though the theory applies to the evolution of all forms of life.

Variations   No two animals are ever exactly alike. Small changes are always appearing in a species as animals produce their young. There are *variations* in colour, shape, weight, height and countless other features.

Now and again an animal may be born with a completely new feature – a colour change or a different skin surface for example. Such variations are called *mutations* and they happen because a mistake occurs in the chromosome copying within a cell. A mutation is normally harmful to an animal but just occasionally it may be useful.

Survival of the fittest   Animals are in a constant struggle for survival. They face difficult weather conditions, they may be hunted, and they must compete with each other for what little food is available. Of all the variations which appear in animals, some will help in the struggle for survival and some will not – some animals survive to reproduce but many do not. Nature 'selects' for survival those animals whose variations have survival-value.

Adaptation   As it is the survivors which reproduce, the variations which help an animal to survive are those most likely to be passed on to later generations. For example:

If long legs aid survival, long-legged animals are those most likely to live long enough to produce young – later generations will inherit their long legs.

By this process of natural selection each species evolves so that its members can best cope with their living conditions – they become *adapted* to their environment.

If the animals of one species begin to live and breed in isolated groups, evolutions will take a different course with each group. Several new species will then evolve.

Mutation in a moth.

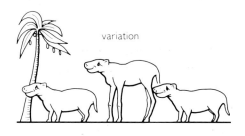

variation

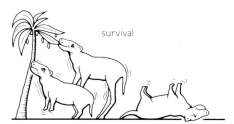

survival

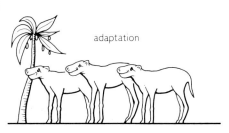

adaptation

Questions

1. Which of the groups of animals on the opposite page evolved from early reptiles?
2. Write down two examples of variations which might be found between animals of the same species.
3. What is a mutation? Give an example.
4. Why are some variations more likely to be passed on to later generations than others?
5. Why may several species evolve from a single species?

Balance of nature

Scientific study of the way in which living things depend upon each other is called ecology. By the end of this section, you will understand why studying it is important.

Food chains

Grass grows by turning carbon dioxide, water and salts into food using energy from the sun. Cows cannot make their food in this way. They eat grass, converting it into their own body materials. We, in turn, drink milk and eat meat to produce our body cells. This is an example of a *food chain*.

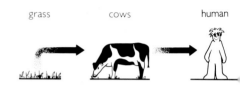

Animals use only one tenth of their food intake for body building. The rest is used for keeping warm, or for keeping moving, or is undigested. The loss at each stage in the food chain is shown by a *pyramid* structure. It takes 1000 tonnes of the microscopic plants to support one tonne of shark through a food chain.

Food webs

Most animals eat more than one kind of food. A *food web* shows how creatures depend upon each other.

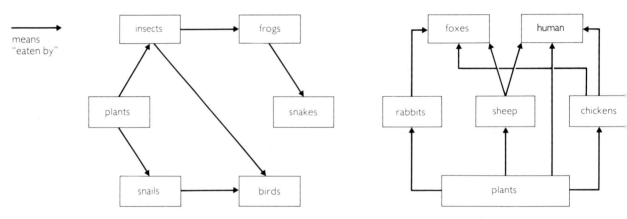

Anything that happens to one member of a food web will affect all the others.

When myxomatosis killed most of the rabbits in Britain, foxes were affected. They ate more domestic animals and reduced one source of our food. Plants were also affected. Seedlings of trees and bushes on moors and grassland began to flourish because there were no rabbits to nip them off!

Predator to prey ratio

There is a natural balance between *predators* (animals which eat others) and their prey. The numbers remain about the same unless something happens to change living conditions, the *environment*.

If, in a stream, too many toads develop, they will eat nearly all the slugs (1). Deprived of food the toads will die off (2). The plants the slugs eat will flourish and the numbers of slugs quickly increase again, with plenty of food and few predators (3). The predator to prey ratio is restored (4).

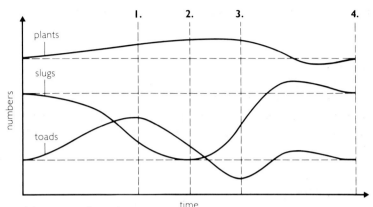

Living together

Life exists where energy from the Sun reacts with water, air and substances from the Earth's crust. This living space is called the *biosphere*. Within the biosphere are many different *habitats* – places where particular things live. Crocodiles, for instance, need the warm muddy banks of slow-moving rivers as their habitat.

All the different things which live in one habitat make up a *community*. Members of a community are often linked by a food web. Sometimes they help each other. Cattle suffer from little creatures called ticks living in their hair. Birds perch on the backs of the cattle eating the ticks. Both birds and cattle benefit from the partnership. It is an example of *symbiosis*.

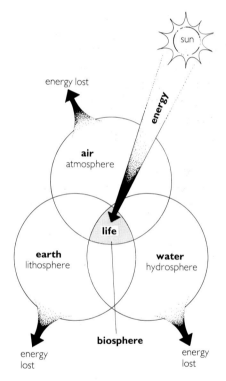

The three sections of a community are:

A **producers** – green plants which make food
B **consumers** – animals which eat the plants (*herbivores*) or each other (*carnivores*)
C **decomposers** – bacteria and fungi which live on dead plants and animals. The dead cells break down into nitrogen compounds and minerals. These collect in the soil and feed more plants.

Thus the community acts upon the environment and the environment acts upon the community. This interaction is called an *ecosystem*.

Questions

1. Give an example of a food chain.
2. What fraction of food is used for body building?
3. Draw the food web between plants, insects, snails and birds.
4. State three different effects of the disease myxomatosis.
5. What is meant by 'predator' and 'environment'?
6. Explain why the predator to prey ratio is roughly constant.
7. Describe the habitat of a crocodile.
8. What are the three sections of a community? What do they do?

Re-using atoms

All living things are made of molecules based on long chains of carbon atoms. Oxygen, hydrogen and other atoms are joined to the carbon chains. As plants and animals grow and die, these atoms are used over and over again.

Essential materials

Plants use energy from sunlight to change carbon dioxide and water into sugar. To make body cells, plants also need large amounts of nitrogen and phosphorus. Calcium and small quantities of about 12 other elements are also needed to go into different kinds of cells. All these elements are absorbed as solutions of mineral salts from soil through the roots of the plants. The weathering of rocks provides a reserve of all the necessary salts except nitrates. These come from the nitrogen in the air.

When animals eat plants, leaves are changed into flesh and blood. The same elements are present but in different combinations. Some of the materials are returned immediately to the soil as droppings. Eventually all the elements become re-usable when the plants and animals die and their bodies decay.

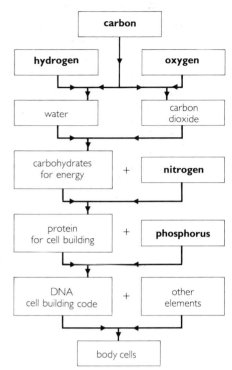

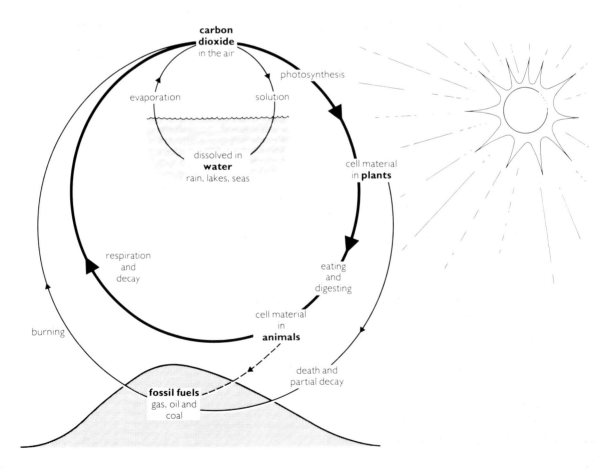

The nitrogen cycle

The nitrogen needed for cell-building proteins also comes from the air. Certain plants are able to absorb nitrogen from the air through lumps on their roots. Such plants are *leguminous*: they include peas, beans and clover. Farmers grow these crops every few years, ploughing parts of them back to enrich the soil with nitrogen compounds. Artificial fertilisers are made by converting nitrogen from the air into ammonium and nitrate salts. These dissolve in water when added to the soil and are absorbed by plants through their roots. Manure (animal excretion) and compost (rotting plants) are rich in nitrogen compounds and these are returned to the soil. Microscopic organisms called *bacteria* play an important part in the recycling of nitrogen.

Clover is often grown to increase the nitrogen content of the soil.

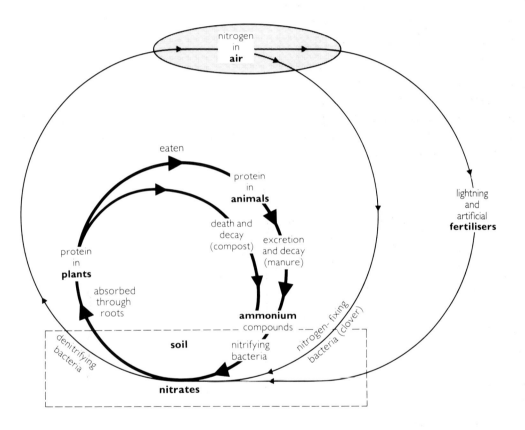

Questions

1. What are the three most abundant elements in living cells?
2. What are proteins? What elements do they contain?
3. What are carbohydrates? What elements do they contain?
4. Where would you find phosphorus in an animal's body?
5. Describe one route for the recycling of carbon atoms.
6. How do all plants absorb nitrogen? Where does the nitrogen come from in the first place?
7. What are leguminous plants? Give two examples.
8. Name two artificial and two natural fertilisers.

Soil

Many of the materials that plants need to produce their cells are taken in through their roots from soil. Since we depend upon plants for our own body cells, soil is vital to us. It's not just dirt!

How soil is formed

Soil comes from broken-down rock. It can form quite quickly, as it did on the volcano which formed the new island of Surtsey in 1964. Plants were growing on this new island within five years.

The actions of wind and rain, sun and ice break rock down into fragments of varying size. Oxygen and carbon dioxide, dissolved in rain water, react with the rock producing clays and soluble salts. These enable small plants to grow. When the plants die, their remains decay to a dark coloured substance called *humus*. It is humus that makes the mixture of rock fragments, clay, water and salts into soil. Humus provides soil with nitrogen and other plant food, and it lightens the soil structure.

Types of soil

No two soils are the same. The structure of a soil depends upon the type of rock it comes from and upon the amount of humus it contains. For good plant growth, soil should crumble easily, be able to hold water with dissolved salts, and have air trapped in it.

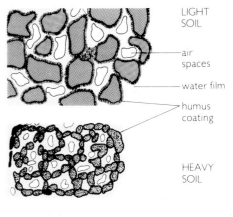

LIGHT SOIL

air spaces

water film

humus coating

HEAVY SOIL

Light soil   This is sandy with large particles giving big air spaces and good drainage. It breaks up very easily when dug and also warms up quickly. Unfortunately, it dries out very quickly and loses its mineral salts. Humus added as farmyard manure holds the sand together and improves its water-holding.

Heavy soil   This consists of small clay particles which stick closely together giving little air space and poor drainage. It is difficult to dig; it becomes water-logged in wet weather; it cracks in dry weather. Addition of humus makes the clay form larger particles (crumbs), allowing more air and water in. Lime has the same effect.

Loam   This is a mixture of sand and clay soils. It has a mixed particle size and rich humus content, which makes it very fertile.

Soil profiles

Fertile soil is usually no more than 20 cm deep. Below it lies the *subsoil* – much lighter in colour than topsoil because very little humus penetrates into it. The layers of soil and subsoil which can be seen in a very deep trench are called the *soil profile*.

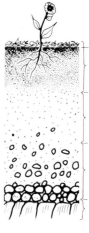

dead leaves

humus + clay, silt, and sand black to grey.

clay + silt, sand and some humus brown to yellow

large stones + some material from layer above

bedrock

Composition of soil

An old can with both ends removed may be used for collecting soil. Pressed into the ground, the can cuts out a sample wthout disturbing the soil structure.

Air   This may be measured by dropping the can into a large cylinder half-filled with water.

Water   Another sample is weighed and then left in a warm place for several days. Reweighing shows the amount of water lost.

Humus   By heating the same dry sample strongly, the humus is burnt away giving further weight loss.

Solid particles   These may be separated by emptying another sample into a cylinder filled with water. When shaken up and left to settle overnight different layers of particles separate.

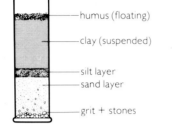

Bacteria in soil

Also present in soil but too small to be seen are microscopic organisms called *bacteria*. Bacteria decompose dead plants and animals forming humus and ammonium compounds. Other *nitrifying* bacteria turn ammonium compounds into nitrates using oxygen trapped in the soil. The bacteria gain energy in this process as animals do by respiration. There are others called *denitrifying bacteria* which break up nitrogen compounds and release nitrogen gas into the air.

Acidity (pH) of soil

The acidity of a soil depends upon the parent rock and the condition of the plant remains. Limestone soils are usually alkaline, pH = 8. Clay soils and bogs with partly-decomposed plant remains (peat) may be strongly acidic, pH = 4 or 5. Most plants grow best in soils of pH = 6.5. Lime counteracts excess acidity and ammonium sulphate can be used to acidify an alkaline soil.

Questions

1. What does soil come from? Describe how it forms.
2. What is the dark substance in soil? How is it formed?
3. What are the requirements of a good soil?
4. Describe light and heavy soils. How does humus improve both?
5. What is the usual depth of soil? What lies beneath it?
6. How can you show that soil contains (a) air, (b) water, (c) humus, (d) sand and silt?
7. Give two effects of bacteria present in soil.
8. Give the range of soil pH. What value suits most plants?

Conservation

The human population is increasing rapidly. There is a constant need for more food crops, more meat, more wood, more fuels and more minerals. We should take care to use our resources wisely.

Problems

Deforestation   Trees are vital to the balance of nature. They play an important part in the carbon cycle; they cause rich soil to develop and they shelter all kinds of small plants and animals. However, people have always destroyed trees for timber, for paper, to make space for farms and towns or just to fight wars more easily! When the trees are gone, soil soon loses its goodness and is often *eroded* – washed away by rain or blown away by wind. The land can turn into desert in hot countries or into barren moors and bogs in cold, wet places.

In spite of these dangers, vast areas of forest in tropical places are being cleared today. A space the size of a football pitch is cleared every second! The land supports crops or cattle for only a few years, then it is abandoned. Some kinds of wild plants and animals may become *extinct* – die out.

Monoculture   is the growing of single crops in very large fields. It produces food cheaply but it allows pests to flourish. Wildlife suffers from the removal of hedges to make the large fields. The ploughing up of down-land in southern England led to the loss of the Large Blue butterfly.

Eutrophication   is the abundant growth of choking weeds and algae when chemical fertiliser is washed into lakes. As the plants die and decay they use up all the oxygen in the water and so all the fish die.

Pesticides   are mildly poisonous chemicals used to kill crop pests and insects. They can kill birds too, as DDT did when it was used on the gnats in Clear Lake, California. A little of the poison was absorbed by plankton; it built up in fish eating the plankton, and it then became finally strong enough to kill birds feeding on the fish.

Over-fishing   Catching all the fish in one place defeats its own purpose. The fish face extinction and the fishermen have nothing left to catch. Today, few sardines are caught off California or herrings around Britian. Similarly, whales have been killed in such large numbers that some kinds, such as the Blue whale, may disappear.

Derelict land   The mining of coal and minerals has left the countryside spoilt by large holes and hills of waste material. These are depressing, dangerous and waste valuable land.

Forests like this . . .

. . . are being destroyed every day.

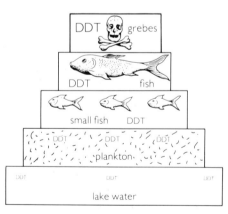

Answers

Farming methods   The need to use chemical fertilisers can be kept to a minimum by growing different crops in rotation over a three year cycle. Seeds of the next crop should be planted as soon as the first crop is harvested, so that the soil is never left bare. Bare soil is avoided by deep ploughing. This destroys weed roots without turning the soil over. Other ways of controlling soil erosion are shown below.

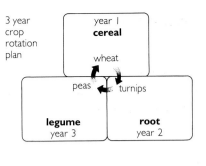

3 year crop rotation plan

year 1 **cereal** — wheat — turnips — peas
legume year 3 — **root** year 2

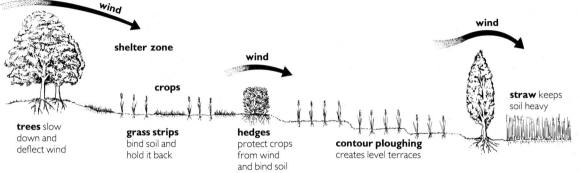

wind

shelter zone

crops

wind

wind

straw keeps soil heavy

trees slow down and deflect wind

grass strips bind soil and hold it back

hedges protect crops from wind and bind soil

contour ploughing creates level terraces

Reforestation   Many high moors in Britain are being made productive again by planting trees. At first, small firs which grow easily are planted close together. As they grow, they are thinned out to create sheltered spots. Here, larger trees will flourish and an undergrowth of shrubs and herbs will develop.

Aqua-farming   Oysters have been bred and grown in big tanks of sea water on French coasts for a long time. It is easy to gather the full-grown shellfish and to control their numbers. Other fish, such as eels and trout, are now being farmed in the same efficient way.

Reclamation   Britain's slag heaps and old quarries are being reclaimed as useful land. Some quarries have been made into lakes for water-sports. Others have been filled with refuse and then grassed over to make parks. Grassed-over slag heaps can be used as sheep farms.

Re-cycling waste   Rubbish contains many materials which can be processed and used again if they are collected separately. Bottle banks for glass are now common. Paper and iron can also be re-cycled easily. This means that the consumption of wood pulp and iron ore can be reduced.

Farming fish.

Questions

1. State four reasons why trees are cut down.
2. What is 'monoculture'? State two problems that it causes.
3. What is DDT? Describe one harmful effect of using it.
4. Give an example of a three year crop-rotation cycle.
5. What is 'erosion'? State four ways of controlling soil erosion.
6. Explain how barren moorlands can be reforested.
7. Name three materials which can be re-cycled profitably.

This abandoned quarry has been turned into a nature park.

179

Pollution

Clean air to breathe, clean water to drink and good soil for growing plants are vital to human life. How strange that we spoil all these things by putting harmful waste substances into them!

Air

We need to burn coal and oil products in homes, factories, cars, ships and aircraft to provide energy. The burning also produces smoke and the poisonous gases carbon monoxide, sulphur dioxide and nitrogen oxides. These can pollute the air very badly. Flourides, hydrocarbons and poisonous metals – especially lead – are other harmful substances put into the air by factories and cars.

Smoke is unburnt carbon (soot) plus tar and hydrocarbons. It blackens buildings, harms crops and can cause lung disease. When smoke is trapped by natural mist or fog, it becomes dangerous yellow *smog*. There was a particularly bad smog in London in the winter of 1952 when 4000 people died from chest complaints. After this, smokeless zones were set up and now smog is rare in Britain.

Carbon dioxide is the main product of burning. It is not poisonous to humans. However it absorbs heat and traps rays from the Sun rather like a greenhouse. If too much carbon dioxide gets into the air, the whole atmosphere will warm up. This 'greenhouse effect' could eventually melt some of the polar ice-caps causing flooding all over the world.

This statue has been damaged by acid rain.

Acid rain The acidic oxide gases of sulphur and nitrogen can cause lung illnesses, such as bronchitis, in industrial areas. They also eat away stone, metal and paint on buildings.

Sometimes the gases rise high into the air and are carried hundreds of miles. They dissolve in the water droplets of clouds and finally fall as acid rain. Factories in Britain are probably responsible for the acid rain in Norway and Sweden. Some lakes in Norway have collected so much acid that the fish are dying. Acid rain is also blamed for killing over half the trees in the forests of Germany and Switzerland.

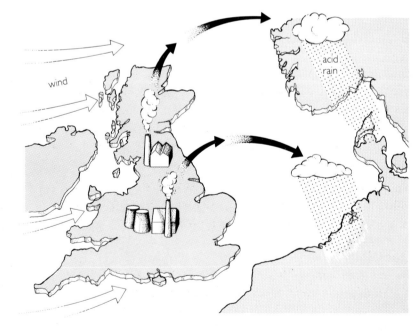

Water

Rivers and lakes can cope with small amounts of pollution from dead plants, fish, and animal droppings. The waste is broken down into harmless substances by bacteria and oxygen present in the water. The waste is said to have a biochemical oxygen demand (BOD); which is the amount of oxygen required to remove it. Water absorbs oxygen from the air very slowly and in small amounts. Large quantities of waste soon use up all this absorbed oxygen. Fish die and bacteria which work without oxygen take over and produce methane (marsh gas) and the foul-smelling, poisonous hydrogen sulphide. To avoid this, untreated sewage is no longer put into rivers. However, it *is* put into the sea in some places. Although it slowly decomposes, sometimes it is washed ashore and fouls the beaches.

Harmful substances can also come from rubbish tips and careless use of pest killers and fertilisers. Factories contribute to water pollution by passing the water they use (*effluent*) back into rivers. In spite of controls, this water sometimes has poisonous chemicals in it or is too hot.

Cars, ships and aircraft give out small amounts of oil all the time. It ends up in rivers or the sea. Occasionally, huge amounts of oil are spilled into the sea from tankers. This oil kills fish and water-birds and fouls beaches.

People cause pollution.

A victim of pollution.

Noise

Noise is unwanted sound. Too much noise can make peope irritable, ill or deaf. Deafness was very common in people who worked in noisy places. Now people using drills and noisy machinery have to wear protectors to save their ears. In spite of the known risk, some people damage their hearing by listening to pop music which is too loud.

Noise can be checked by instruments which record the level in units called *decibels* (db). Anything over 80 db becomes uncomfortable. Airport noise at 140 db is a serious problem.

Questions

1. List five harmful substances produced by a car engine.
2. What is the 'greenhouse effect'? What problem could it cause?
3. What is 'acid rain'? State two problems caused by acid rain.
4. How do rivers get rid of small amounts of organic pollution?
5. Name the sources of three different kinds of pollution in the sea.
6. What is noise? How can very loud noise affect people?

	decibels
	160
Damage to ear drums	140
Jumbo jet taking off	130
Thunder/motor bike	120
Pop music concert	110
Food mixer	100
Tube train	90
Alarm clock	80
Telephone	70
Normal speech	60
	50
Whispering	40
Watch ticking	30
	20
Peace and quiet	

The human skeleton

There's plenty of soft tissue in the body – but much harder materials are needed to enable you to stand, walk and chew your food!

The skeleton

All vertebrate animals, including human beings, are supported by a framework of rigid bones called a *skeleton* – it is called an *endoskeleton* because it lies inside the body.

The skeleton has several important functions:

Support   The skeleton enables you to stand upright on the ground. It also supports vital internal organs so that there is no danger of them squashing one another.

Protection   The skeleton protects many organs from outside damage:

the *skull* protects the brain;

the *rib cage* protects the heart and lungs;

the *vertebral column* (the backbone) protects the spinal cord.

Movement   Many parts of the skeleton are jointed so that body movements are possible; e.g. arms, legs, fingers and jaw.

The movements are produced by muscles which surround the skeleton and are attached to it at various points.

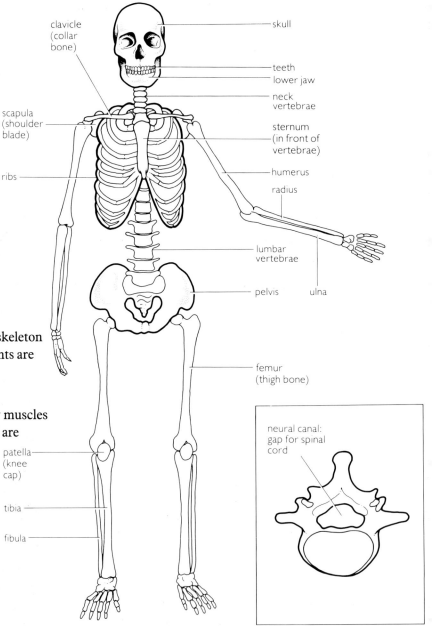

clavicle (collar bone)
scapula (shoulder blade)
ribs
skull
teeth
lower jaw
neck vertebrae
sternum (in front of vertebrae)
humerus
radius
lumbar vertebrae
pelvis
ulna
femur (thigh bone)
patella (knee cap)
tibia
fibula

neural canal: gap for spinal cord

Lumbar vertebra (top view).

182

Bones

Bones contain living cells surrounded by hard minerals to give strength and rigidity – calcium is the main substance present. These minerals are constantly being renewed from materials brought in by the body's blood supply; if this did not happen, broken bones would never be able to mend.

Bones are made even stronger by tough fibres which reinforce them in the same way as steel rods reinforce concrete.

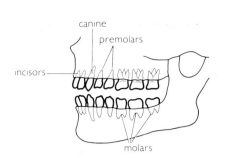

Teeth

Human adults have 32 permanent teeth in all. Your permanent teeth probably began to replace your 20 milk teeth when you were about six years old.

Incisors are cutting teeth. You use them when you bite a sandwich.

Canines are gripping and tearing teeth. Flesh-eating animals like dogs have well-developed canine teeth. Our own canine teeth are much smaller in comparison.

Premolars and molars are grinding and crushing teeth. They are used to break food down into small pieces.

Tough fibres hold each tooth in its socket in the jaw bone; they cushion the tooth against jolts during chewing.

Enamel covers the crown of each tooth. It gives the tooth a hard surface for cutting and grinding.

Dentine is similar to bone in structure. Most of the hard material in a tooth is dentine.

Pulp at the centre of each tooth is made of soft tissue containing blood vessels and nerves.

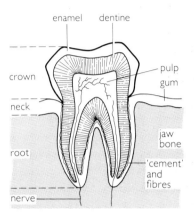

Molars: crushing teeth.

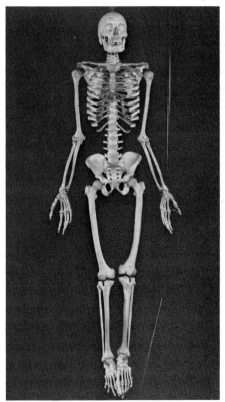

How many of the bones in this skeleton can you identify?

Questions

1. Which parts of the human body are protected by: (a) the skull; (b) the rib cage; (c) the vertebral column?
2. What other functions does the skeleton have?
3. Draw labelled diagrams to show which limbs of the body contain
   (a) a femur, tibia and fibula;
   (b) a humerus, radius and ulna.
4. What is the main substance present in bone?
5. Which teeth do you use to cut through food?
6. Which teeth do you use to crush food into small pieces?
7. What covers the crown of each tooth, and why?

Joints and muscles

Joints are places in the skeleton where bones meet. There's movement at most joints, caused by muscles such as 'bulging biceps'.

Joints

At some joints, the bones are held tightly together by fibres and no movement is possible – the bones which form the top of the skull are joined in this way. Other joints allow small movements because flexible *cartilage* (gristle) is sandwiched between the ends of the bones. Your back can bend a little because the vertebrae have cartilage *discs* between them which absorb the jolts when you walk about.

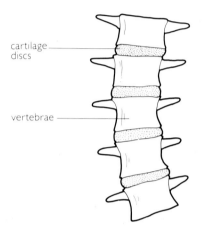

Most joints can move much more freely. Tough fibres called *ligaments* hold the ends of the bones together while allowing the bones themselves to be moved through large angles. These joints are called *synovial* joints. They include:

Hinge joints   Knuckles, knees and elbows are all hinge joints. In hinge joints, the bones can be turned to and fro in one direction only – rather like a door on its hinge.

Ball-and-socket joints   Hip joints and shoulder joints are of this type. In ball-and-socket joints, the bones can be turned in any direction – up and down or side to side.

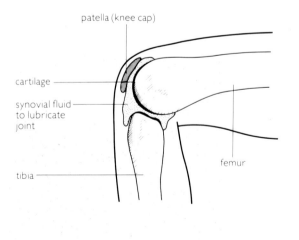

A hinge-joint: the knee.

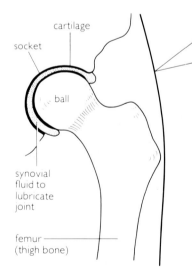

A ball-and-socket joint: the hip

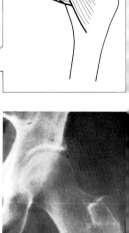

X-ray picture of a hip joint

Muscles

Joints are moved by muscles which are attached to the bones by strong, non-stretch fibres called *tendons*. The muscles pull on the bones by thickening and contracting (becoming shorter) – this happens whenever impulses are received from the body's nervous system.

The diagram on the right shows the muscles which move the forearm up and down. Like the other muscles attached to the skeleton, these are *voluntary* muscles – you are free to move them whenever you choose.

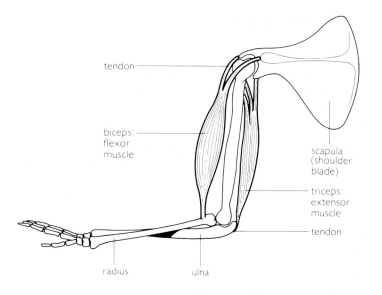

Antagonistic muscles   After contracting, a muscle cannot lengthen itself again; it has to be pulled back to its original shape. For this reason, muscles are often arranged in *antagonistic pairs* with one pulling one way and the other pulling back. The muscles which move the forearm are arranged in this way – the *flexor* muscle bends the joint, the *extensor* muscle straightens it again.

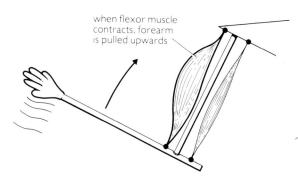

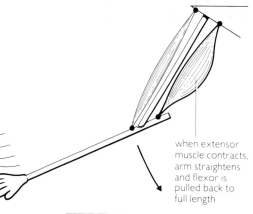

Questions

1. Why is your back able to bend a little?
2. In what way is a ball-and-socket joint different from a hinge joint? Give two examples of each. How are these joints lubricated?
3. What are   (a) ligaments;   (b) tendons? What do they do?
4. What happens to a muscle when it receives an impulse from the nervous system? What is a voluntary muscle?
5. Why are muscles often arranged in antagonistic pairs?
6. What is the difference between a flexor muscle and an extensor muscle? Give an example of each.

A flexor demonstration.

Sense and control

What makes you move your arm, or breathe more quickly, or grow? Something somewhere has a great many decisions to make, some of which your know about and some of which you don't.

The nervous system
The body is controlled by the *brain* and *spinal cord*. Together these are known as the *central nervous system* and they are linked to all parts of the body by *nerves*. Through the nerves, the central nervous system can sense what is happening inside and outside of the body and can control the actions of the muscles and organs.

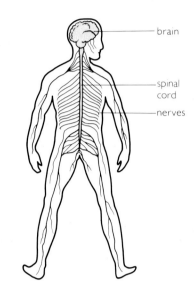

Sense organs
The central nervous system gets its information from tiny sense organs or *receptors* which lie at the ends of the nerves. These send out small electrical impulses which are passed back to the central nervous system through chains of nerve cells. Some sense organs check what is happening in organs in the body, others give you the senses of sight, sound, balance, taste, smell and touch:

Sight, sound and balance   The eye and the ear are described on pages 94 and 104. The inner ear contains the organs of balance.

Taste and smell   Your tongue is covered with tiny bumps called *papillae*. The sense organs are called *taste buds* and they lie on the side of each papilla. There are four types of taste bud; they are concentrated in different areas of the tongue and they each respond to one of the tastes *sweet*, *sour*, *salt* or *bitter*. Most tastes are a mixture of these different sensations.

The *olfactory organs* within the nose are the sense organs of smell. During eating, these sense organs respond to vapours from the food. Much of your sense of taste is really a sense of smell.

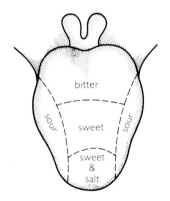

Taste areas of the tongue.

Touch   The sense organs of touch lie in or just under the skin. Different types of sense organ enable you to sense touch, pressure, hot or cold, pain, and hair movements.

The skin itself is made up of two layers, the *dermis* and the *epidermis*, underneath which are layers of fat. Skin is the body's protective coating. It is also used in keeping the body at a steady temperature.

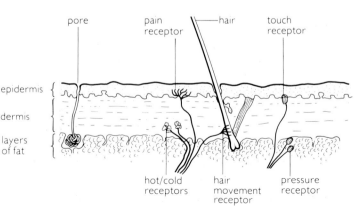

Control through the nervous system

Acting on the information it receives, the central nervous system sends out its instructions through chains of *motor* nerve cells. Some instructions you have control over and some you do not – you can decide when to contract muscles to move a leg but not when to make your heart beat faster.

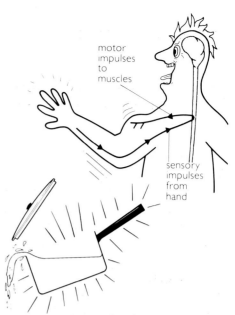

motor impulses to muscles

sensory impulses from hand

Reflex actions   Movements or changes in the body which take place automatically are called reflex actions. They include coughing, sneezing and blinking, and the many actions caused by the *autonomic nervous system*.

The autonomic nervous system   is the part of the central nervous system that controls the internal workings of the body such as:

speeding up or slowing down the heart beat;
control of body temperature;
control of breathing rate.

Not all reflex actions involve the brain. It is a *spinal reflex* which pulls your hand away from a hot saucepan if you touch it accidentally – the motor nerve impulses are sent to the arm muscles straight from the spinal cord.

The endocrine system

The central nervous system has second-by-second control of the body. The endocrine system controls much slower bodily changes. It consists of a number of *glands* which release chemicals called *hormones* into the blood stream. These hormones affect the way in which the organs of the body do their work, controlling amongst other things:

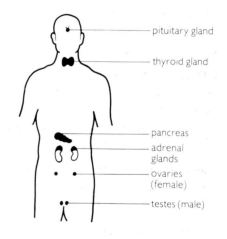

pituitary gland

thyroid gland

pancreas

adrenal glands

ovaries (female)

testes (male)

the rate at which you grow;
the development of some sexual characteristics;
the control and use of sugar in the blood.

The *pituitary gland* releases a growth hormone. It also releases hormones which control the other endocrine glands. For this reason, it is sometimes called the 'master gland'.

Questions
1. Which parts of the body form the central nervous system?
2. What is a receptor? Give three examples of sense organs.
3. What taste sensations are detected on the tongue?
4. Some nerve impulses travel in to the central nervous system and some travel out. Why?
5. What are the two main skin layers known as?
6. Give three examples of reflex actions.
7. What is the autonomic nervous system? What does it control?
8. Where do hormones come from and what do they do?
9. Why is the pituitary gland called the 'master gland'?

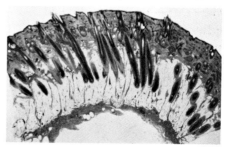

Cross section through the skin.

Foods

To keep alive, the body needs to take in water, oxygen from the air, and food. But it needs more than one type of food.

Nutrients

There are five main types of food or *nutrient:*

Carbohydrates and fats   supply you with most of your energy.

Proteins   provide the materials from which new body tissues are built when you grow or when, for example, a wound heals.

Minerals   are used in building some body tissues. They are also needed for some chemical reactions in the body.

Vitamins   are vital for speeding up some chemical reactions in the body that might otherwise happen too slowly.

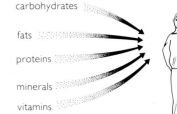

Most foods contain several of the nutrients described above. A balanced diet is one which supplies you with adequate amounts of *all* the nutrients. The table on the next page gives *some* of the foods which are a common source of each type of nutrient. Below, are ways of testing for these nutrients.

Food tests

Foods can be tested to discover which nutrients they contain. Some simple examples are given below:

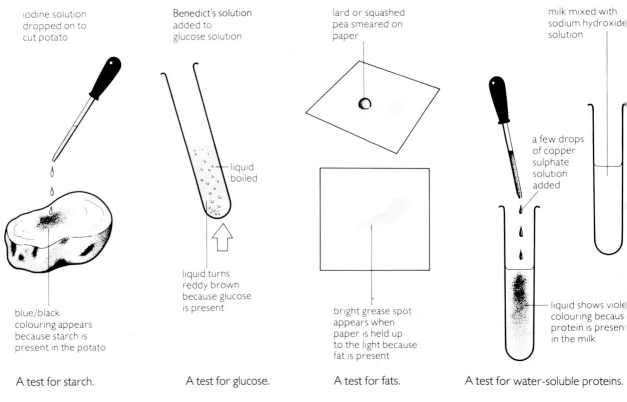

iodine solution dropped on to cut potato	Benedict's solution added to glucose solution	lard or squashed pea smeared on paper	milk mixed with sodium hydroxide solution

liquid boiled

a few drops of copper sulphate solution added

liquid turns reddy brown because glucose is present

blue/black colouring appears because starch is present in the potato

bright grease spot appears when paper is held up to the light because fat is present

liquid shows viole colouring becaus protein is presen in the milk

A test for starch.

A test for glucose.

A test for fats.

A test for water-soluble proteins.

Carbohydrates	*Sugars*	*Starches*	*Cellulose*
	Jam, cakes, sweets, glucose, sweet fruits	Potatoes, rice, bread, flour	Vegetables, cereal foods
	The body breaks down sugars and starches into simple sugars like glucose. Some may be converted into fats.		*Cellulose cannot be used by the body as a nutrient, but it provides bulk (dietary fibre or 'roughage') to help food pass through the system more easily.*

Fats	Butter, margarine, lard, meat.
	Fats can be stored by the body; they provide a reserve supply of food.

Proteins	Meat, eggs, fish, milk, cheese, bread
	The body breaks down proteins into amino acids which it can use to build new body tissues

Minerals	Minerals needed by the body include:
	Calcium *(for teeth and bones)* – from cheese, milk, vegetables
	Iron *(used in making blood)* – from liver, eggs, bread
	Sodium *(for muscle movements)* – from salt

Vitamins	Vitamins needed by the body include:
	Vitamin A – from green vegetables, carrots, liver, butter.
	(a shortage of vitamin A weakens your vision in the dark)
	Vitamin B$_1$ (thiamine) – from yeast, bread, meat, potatoes, milk.
	Vitamin B$_2$ (riboflavine) – from fresh milk, liver, eggs.
	Vitamin C – from blackcurrants, green vegetables, oranges.
	(a shortage of vitamin C causes a disease called scurvy)
	Vitamin D – cod liver oil, margarine, eggs.
	(a shortage of vitamin D causes rickets – soft bones.
	Your skin makes vitamin D when exposed to sunlight.)

Questions

1. Which type of food provides materials for body building? Name two foods containing this type of nutrient.
2. Which carbohydrate supplies dietary fibre? Why is it useful?
3. Which nutrient can we store in the body as reserve food?
4. Name three minerals which are needed by the body.
5. Which vitamins are needed to prevent scurvy and rickets?
6. What foods are useful sources of vitamin C?
7. Write down all the nutrients given in the tables which are supplied by (a) bread (b) milk (c) cheese (d) liver.
   Your list will not show all the nutrients in each food.

Digesting and absorbing food

Swallow food, and the useful substances eventually end up in the blood stream to be carried to wherever in the body they are needed.

The alimentary canal

When you swallow food, it starts to move down a long tube called the *alimentary canal*. This tube runs from the mouth to the anus and food is squeezed along it by wave-like muscle movements called *peristalsis*.

Two important things happen to the food as it passes along the alimentary canal:

The food is digested   During digestion, chemical compounds called *enzymes* break the food down into simpler substances which can be carried in the bloodstream, and which will dissolve in the water within the cells of the body.

Digestion mainly takes place in the stomach and the small intestine, but it starts in the mouth – an enzyme in saliva starts to break down starch into glucose as you chew your food.

Digested food is absorbed into the blood stream   This mainly takes place in the small intestine – its walls are lined with tiny blood tubes which carry the digested food away.

Undigested matter passes into the large intestine and is pushed out of the anus when the toilet is used.

What happens to food in each part of the alimentary canal is described in more detail on the opposite page.

mouth

tongue

salivary gland

trachea (wind pipe)

swallow food

gullet

stomach

gall bladder

liver

pancreas

duodenum

small instestine

ileum

colon

large instestine

rectum

appendix

anus

The stomach

After food has been swallowed, it passes down the gullet and into the stomach. Here, muscle movements in the stomach walls mix the food with *gastric juice* which oozes from thousands of tiny glands in the stomach lining. Gastric juice contains an enzyme which starts to digest any proteins in the food. It also contains dilute hydrochloric acid which the enzyme needs in order to work properly.

The stomach stretches as it fills with food. Every now and then, a ring of muscles at the bottom of the stomach allows some mixed food to pass down into the small intestine.

The small intestine

In the upper part of the small intestine, two more liquids are mixed with the food to help the process of digestion:

Bile, which is made in the liver and stored in the gall bladder, breaks up fats into droplets small enough to be digested.

Pancreatic juice from the pancreas carries powerful enzymes which digest starches, fats and proteins. It also neutralises the stomach acid.

Digestion is completed in the small intestine and digested food passes into the blood through millions of tiny bumps on the intestine walls. The bumps are called *villi* – they give the small intestine a large surface through which to absorb food, and they contain the network of tiny blood tubes which carry the food away.

Undigested droplets of fat are also absorbed by the villi. They pass into tiny tubes containing a liquid called *lymph* which drains into the blood elsewhere in the body.

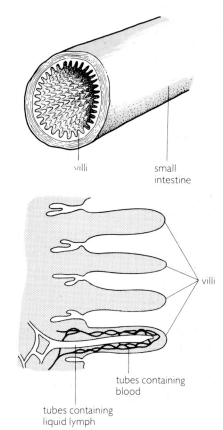

villi   small intestine

villi

tubes containing blood

tubes containing liquid lymph

The large intestine

Undigested matter such as cellulose passes into the large intestine. Here, water remaining from the digestive juices is absorbed so that it can be re-used by the body. Waste matter, now semi-solid, passes out of the body through the anus.

Questions

1. Make a list of the main sections of the alimentary canal starting at the mouth.
2. What is peristalsis? What does it do to the food?
3. What is happening to food during digestion?
4. Where does digested food eventually end up?
5. What does gastric juice contain? Why?
6. What jobs are done by   (a) bile;   (b) pancreatic juice?
7. In what part of the alimentary canal is digested food mainly absorbed? Through what is it absorbed?
8. Where in the alimentary canal is water mainly absorbed?

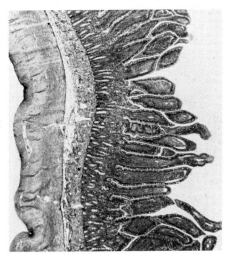

Cross section through the wall of the small intestine.

The blood system

Blood carries digested food to the cells of the body. It also brings them other vital substances such as oxygen and water, and it carries away their waste materials.

Blood

Blood is a mixture of *red cells*, *white cells* and *platelets* all suspended in a watery liquid called *plasma:*

Red cells   These are made in the red marrow inside bones. Red cells are partly made from a protein called *haemoglobin* which contains iron and gives blood its dark red colour. Haemoglobin is important because it can attract and carry molecules of oxygen. Blood which is carrying oxygen is bright red in colour.

White cells   There are fewer white cells in the blood than red, and they are not all of the same type. Some white cells surround and digest harmful germs and dead cells, others make chemicals called *antibodies* which help you fight off disease.

Platelets   These are pieces of broken up blood cells which come from red bone marrow. They help the blood to clot when the skin is cut.

Plasma   Plasma is mainly water but it contains many dissolved substances including digested foods, hormones, antibodies, and waste materials such as carbon dioxide.

Circulation and the blood

The heart pumps blood round the body through a system of tubes:

Arteries   Blood leaves the heart through wide tubes or *vessels* called arteries. These have thick walls to withstand the high pumping pressure.

Capillaries   The arteries divide into narrower tubes which carry the blood to networks of very fine tubes called capillaries. Every living cell in the body lies close to a capillary – water from the blood seeps out through the thin capillary walls bringing dissolved food and oxygen to the nearby cells. It also carries their waste materials back into the capillary. The watery liquid surrounding each cell is called *tissue fluid.*

Veins   Blood from the capillaries drains into wider tubes called veins and returns to the heart at low pressure. Some veins, such as those in the legs, contain one-way valves to stop the blood flowing backwards.

cross section

Red blood cells.

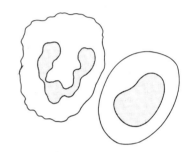

White blood cells.

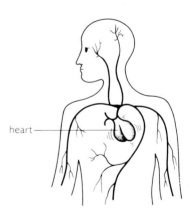

Platelets.

heart

The heart

The heart is really two separate pumps in one. One pump sends blood through capillaries in the lungs, where oxygen is absorbed; the other pump takes in this oxygen-carrying blood and pumps it round the rest of the body. This second pump is made of thicker tissue than the first because it has to pump blood through a much longer system of blood tubes.

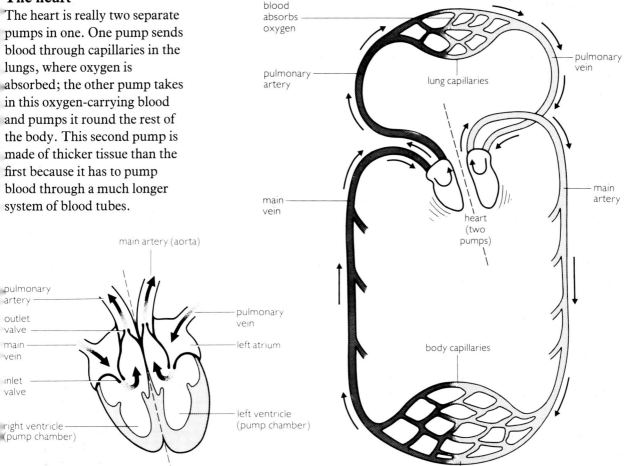

Double circulation of the blood.

Each pump has two valves. These let blood through only in one direction. Between the valves is a chamber called a *ventricle*. The chamber is made smaller when the muscles around it contract, and blood is pushed out through the outlet valve. When the muscles relax again, more blood flows into the chamber through the inlet valve. The muscle contractions or 'beats' are set off by nerve impulses produced in the heart itself, though the beat-rate can be changed by the autonomic nervous system – your heart beats at anything from 60 to 150 times a minute depending on what you are doing.

Questions

1. Which type of blood cell contains haemoglobin? Why is haemoglobin important?
2. Why is some blood dark red in colour and some bright red?
3. What does blood plasma contain?
4. What are the tubes called through which blood (a) leaves the heart (b) returns to the heart?
5. What is (a) a capillary; (b) tissue fluid?
6. Where does blood pumped from the right ventricle travel?
7. What causes the muscles in the heart to contract?

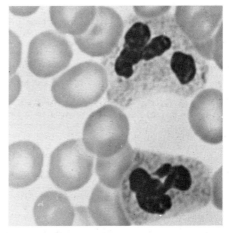

Human blood, showing red and white cells.

The lungs

In the lungs, oxygen is added to the blood and carbon dioxide is removed. It's all part of the process by which the body releases the energy in digested food.

Respiration

The cells of the body get the energy they need by combining digested food with oxygen brought to them by the blood. The process is called *tissue respiration* – like the burning of other fuels, it produces water and carbon dioxide gas.

As blood passes through the lungs, it absorbs fresh supplies of oxygen from the air, and releases the unwanted carbon dioxide carried away from the cells.

The lungs

The lungs are two spongy bags of tissue filled with millions of tiny air spaces called *alveoli*. These have very thin walls and are each surrounded by a dense network of fine blood tubes (capillaries). Oxygen from the air you breathe diffuses through the thin walls and into the blood; at the same time, carbon dioxide diffuses from blood out into the air spaces.

The tiny air spaces lie at the ends of a series of branching air passages which connect with the *windpipe*. As you breathe, air moves in and out of the lungs through the windpipe and some of the old air in the air spaces is exchanged for new.

lime water turns milky

Carbon dioxide is present in the air you breathe out.

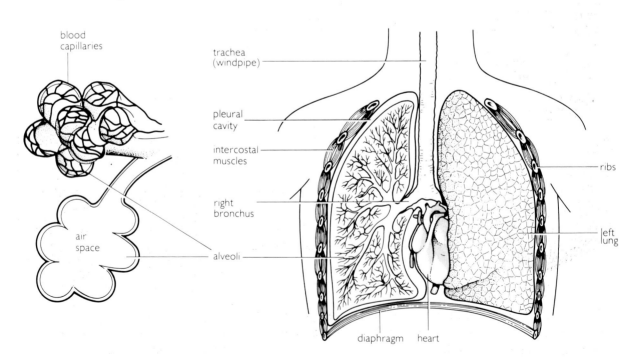

blood capillaries

air space

trachea (windpipe)

pleural cavity

intercostal muscles

right bronchus

alveoli

ribs

left lung

diaphragm   heart

Breathing

The lungs hang in an airtight space called the *thorax*. At the bottom of the thorax is a curved sheet of muscle called the *diaphragm*. Around the sides of the thorax is the rib-cage, with sheets of muscles between the ribs – these are known as *intercostal muscles*.

Breathing in When you breath in, two things happen at once; the diaphragm muscles pull the diaphragm downwards and the muscles between the ribs pull the rib-cage upwards and outwards. These muscle movements increase the size of the thorax, and the lungs expand as air flows in to fill up the extra space.

Breathing out When you breathe out, the diaphragm muscles and the muscles between the ribs relax, the thorax returns to its original size and air is pushed out of the lungs.

During light breathing, the breathing action is mainly produced by movement of the diaphragm muscles. The muscles between the ribs are used when greater amounts of air are moved in and out of the lungs – during running for example. With more air flowing in and out of the lungs, the blood can absorb oxygen and release carbon dioxide at faster rates.

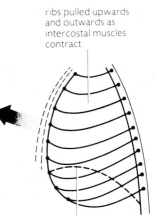

ribs pulled upwards and outwards as intercostal muscles contract

diaphragm

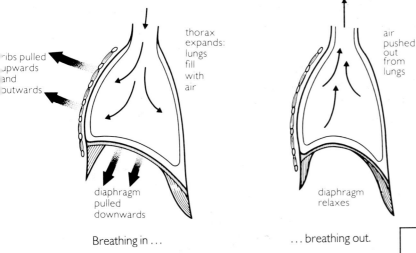

thorax expands: lungs fill with air

ribs pulled upwards and outwards

diaphragm pulled downwards

Breathing in . . .

air pushed out from lungs

diaphragm relaxes

. . . breathing out.

Questions

1. In your lungs, what gas is taken into the blood? How do the cells of the body make use of this gas?
2. In your lungs, what gas is removed from your blood? What simple test could you carry out to show that this gas is present in the air you breathe out?
3. What are alveoli? Why are they surrounded by capillaries?
4. What is (a) the windpipe; (b) the thorax?
5. What two sets of muscles expand your thorax when you breathe in? What effect does each set of muscles have? Which muscles are used during light breathing?

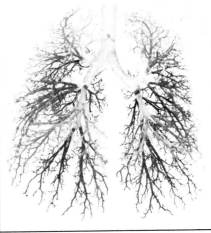

The air system of the lungs.

195

Balance in the blood

Blood has many different jobs to do as it circulates round the body, but the various substances it carries must all be kept in the right proportions.

Functions of the blood

Blood does a variety of different jobs:

It carries oxygen and digested foods such as glucose to the cells of the body;
It carries carbon dioxide and other waste materials away from the cells;
It carries hormones around the body;
It distributes heat energy throughout the body.

The different substances carried by the blood must all be kept in the right proportions. This job is done by the liver, the lungs and the kidneys as blood circulates through them.

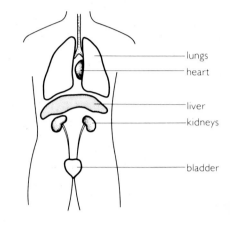

The liver

The liver is a very complex chemical factory – the diagram below shows just some of the processes which take place within it. It constantly adjusts the concentrations of many substances in the blood – glucose is one example; it stores some substances and chemically changes some to make them more useful or less harmful. Most of the digested food absorbed in the small intestine is carried by the blood straight to the liver.

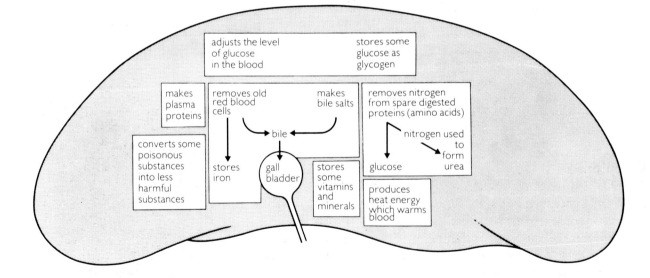

The varied functions of the liver.

Organs of excretion

The lungs and the kidneys are organs of excretion – they remove from the body substances which it makes but does not need.

The lungs These remove carbon dioxide from the blood. They also add oxygen to the blood.

The kidneys The kidneys remove from the blood nitrogen-rich compounds such as urea. They also remove spare salts and water which the body has no use for. Together, these substances form urine which is collected in the bladder and released when the toilet is used.

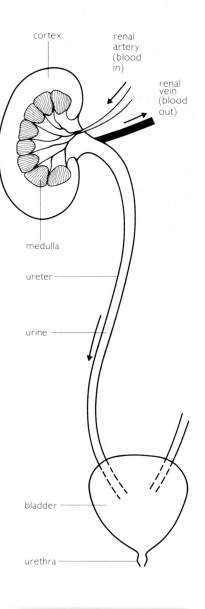

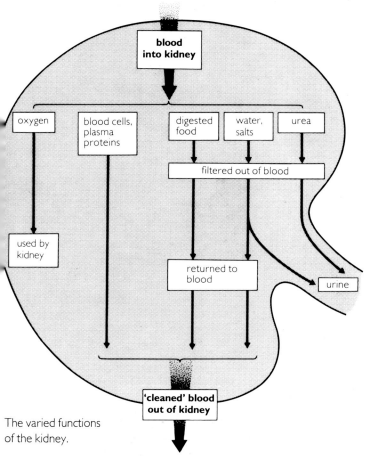

The varied functions of the kidney.

Questions

1. What jobs are done by the blood as it circulates?
2. What happens to most of the digested food absorbed in the small intestine? Name a digested food.
3. Which organ adjusts the concentration of glucose in the blood?
4. What does the liver do with spare digested proteins (amino acids) which the body has no use for?
5. What happens to urea produced in the liver?
6. Write down three substances stored in the liver.
7. Which organ excretes   (a) carbon dioxide;   (b) urine?
8. What does urine contain? Where in the body is it collected?

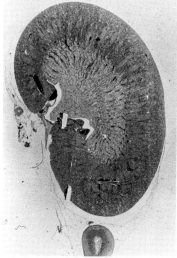

Cross section through a human kidney.

197

Reproduction (1)

A human baby grows from a tiny cell within its mother. This cell is formed when an egg inside the mother is fertilised by a sperm from the father.

Puberty
Puberty is the start of the time when a girl is able to become a mother and a boy to become a father. Puberty often occurs around the ages twelve to fourteen, but it is quite normal for it to be earlier or later than this.

The female reproductive system
At birth, a girl has hundreds of thousands of partly developed eggs or *ova* in her *ovaries*. From the time she reaches puberty, she releases a mature egg from one of her ovaries about every 28 days. The egg is drawn down the egg tube and into the *uterus* (the womb).

The release of an egg is called *ovulation*. Just before this happens, the ovaries put into the blood a hormone called *oestrogen* which sets off changes in the lining of the uterus – the lining thickens and a dense network of tiny blood tubes forms within it. The uterus is then ready to receive a fertilised egg and to nourish it as it develops into a baby.

Menstruation   If the egg is not fertilised, it dies and the lining of the uterus begins to break up. Within about two weeks, ovum, lining and blood pass out through the vagina during a period – the process is called *menstruation*.

The 28 day cycle of ovulation, uterus lining growth and menstruation is called the *menstrual cycle*.

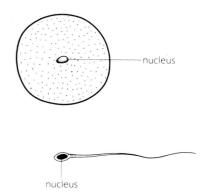

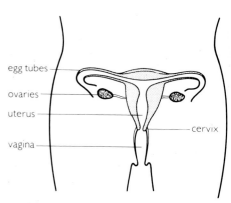

A human egg (or *ovum*) and a human sperm – both greatly enlarged.

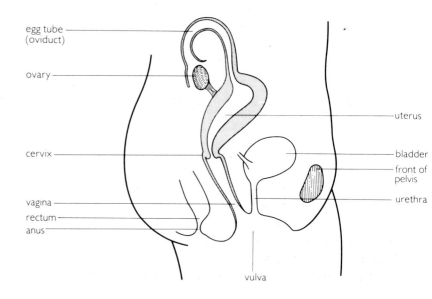

The male reproductive system

A man produces sperm cells in his *testes*. To leave his body, the sperms first travel through the sperm ducts where they mix with liquid produced in the seminal vesicles and the prostate gland.

The liquid and sperms are together known as *semen*. The liquid nourishes the sperms and sets off the swimming movements in their tails which help them to travel up to the egg tubes of the woman. Like urine, semen leaves the body through the urethra, the tube that passes through the penis.

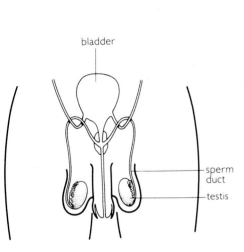

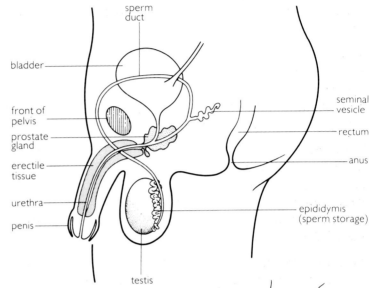

Fertilisation

When a man and a woman have sexual intercourse, the man's penis is placed in the woman's vagina. The vagina and the vulva release fluid which helps entry of the penis, and the penis is stiffened by blood which builds up within its tissue. A reflex action called *ejaculation* then takes place in the man – rhythmic muscle contractions pump semen from the penis.

The small amount of semen released contains millions of sperms, many of which pass into the uterus. Several thousand sperms may reach the egg tubes where they may meet an egg. Despite the high number of sperms present, only one fertilises the egg.

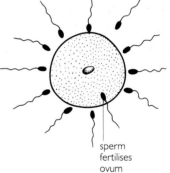

sperm fertilises ovum

Questions

1. Where in a woman are eggs (ova) produced?
2. What is ovulation? How often does it occur?
3. What happens to the lining of the uterus just before ovulation takes place? What then happens to it if an egg is not fertilised? What is this process called?
4. About how long is the menstrual cycle?
5. Where in a man are sperms produced?
6. How many sperms are needed to fertilise an egg?

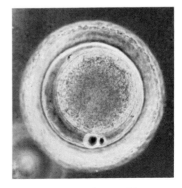

The beginnings of a new life.

199

Reproduction (2)

As a fertilised human egg grows into a baby, it is nourished and protected in the mother's uterus. About nine months after fertilisation, the baby is born.

From egg to embryo

When the egg is fertilised, the head of the sperm is taken into the egg and the tail is left outside. An extra 'skin' now grows round the egg to keep out other sperms. Inside the egg, the nucleus of the sperm joins with the nucleus of the egg, forming a single nucleus which contains full chemical instructions for a baby to be 'built' by cell division.

The fertilised egg divides over and over again as it passes down the egg tube and into the uterus. As more and more cells are produced, they form into a tiny *embryo* which embeds itself in the thick uterus lining.

The growing embryo

After six weeks, the embryo is about as long as a finger-nail. It has a pumping heart and a brain and it lies in a water-filled bag called the *amnion* which protects it from jolts and bumps.

An *umbilical cord* links the growing embryo to a special organ called the *placenta* which has developed on the uterus lining. The embryo's blood circulates through the placenta. As it does so, it absorbs food and oxygen from the mother's blood and releases carbon dioxide and other waste materials into it. Although the two blood systems are entirely separate, they run very close together where the placenta meets the uterus lining, and materials can pass between the two by diffusion.

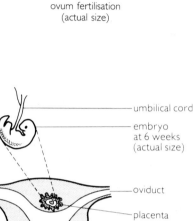

ovum fertilisation
(actual size)

umbilical cord

embryo
at 6 weeks
(actual size)

oviduct

placenta

uterus

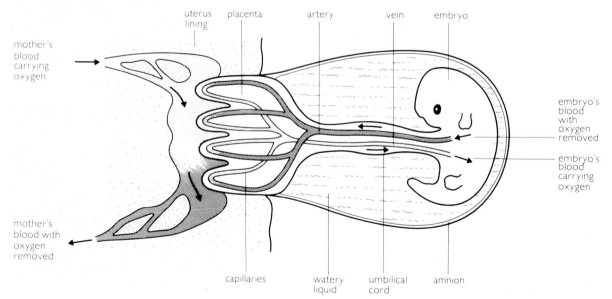

mother's blood carrying oxygen

uterus lining

placenta

artery

vein

embryo

embryo's blood with oxygen removed

embryo's blood carrying oxygen

mother's blood with oxygen removed

capillaries

watery liquid

umbilical cord

amnion

Two months after fertilisation, the embryo is very baby-like in appearance. Now called a *foetus*, it has a face, limbs, fingers and toes.

At six months, it is about 40 centimetres long and quite able to kick its mother.

Birth usually occurs about nine months after fertilisation.

Birth

A few days before birth, the baby turns until its head lies just above the uterus entrance, a ring of muscle called the *cervix*. As birth approaches, muscles which have developed in the walls of the uterus begin to make regular rhythmic contractions and the cervix starts to open. When the opening in the cervix is wide enough, the baby's head passes down into the vagina. At about this time, the amnion bursts and the watery liquid runs out.

Powerful muscle contractions push the baby from the uterus and out of the mother. The same muscles push out the placenta (the 'afterbirth') shortly after the baby is born.

The baby gives a loud cry as its lungs fill with air for the first time, helped sometimes by a slap on the bottom from the midwife. The baby must now take in its own oxygen and food rather than use its mother's. Soon after birth, the umbilical cord is clipped and cut – the remains of the cord shrivel away to leave the navel ('belly button').

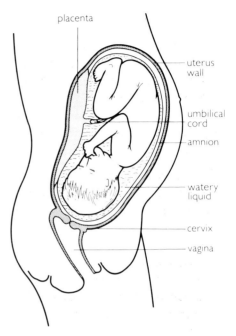

A few days before birth . . .

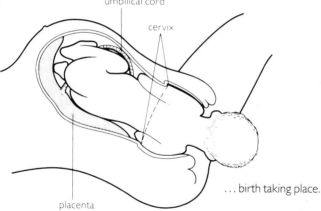

. . . birth taking place.

Questions

1. How does an egg stop the entry of more than one sperm?
2. As cell division takes places, what does a fertilised egg develop into? Where does this embed itself?
3. What is the amnion? What does it contain and why?
4. What joins the embryo to the placenta?
5. Where does food and oxygen pass from the mother's blood into the embryo's blood? What else passes between the blood systems?
6. How many months are there between fertilisation and birth?
7. What happens to the baby a few days before birth?
8. What is the 'afterbirth'?

A new life taking shape.

Germs and diseases

There are vast numbers of tiny living organisms in our bodies and in air, soil, water and plants. Although they are too small to be seen, many of these microbes do useful jobs. Some, however, are harmful to the body and cause disease: these are often called germs.

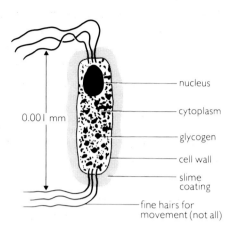

Bacteria

These are very small, living cells about 0.001 mm long. They can only be seen under powerful microscopes. Under suitable conditions, a bacterium may divide every half hour, producing over a million in 10 hours. The bacteria which cause disease either attack the body tissues or release poisons.

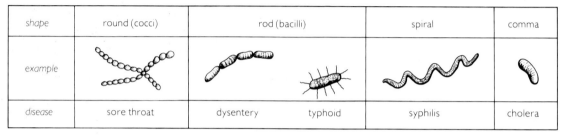

shape	round (cocci)	rod (bacilli)		spiral	comma
example					
disease	sore throat	dysentery	typhoid	syphilis	cholera

Viruses

Viruses are 10 times smaller than bacteria and can pass through filters which trap bacteria. Unlike bacteria, which can also exist in water or soil, viruses can only live and multiply in other living tissues. Viruses cause the common cold, polio, 'flu and mumps.

Fungi

Mushrooms, yeast and the mould which grows on bread are all fungi. Some of the microscopic fungi attack the skin, the mouth and the throat. One of these is ringworm and another is athlete's foot. In 1929 Alexander Fleming found that the *penicillium* mould stopped the growth of bacteria. Penicillin, extracted from the mould, was the first *antibiotic*. Antibiotics are now widely used to control disease.

Spread of disease

There are four main ways in which germs can be picked up.

Droplet infection   Whenever you breathe out, droplets of moisture leave the body. People with diseases such as influenza breathe out germs in these droplets.

Contact   Contagious diseases, e.g. measles, can only be picked up by touching an infected person or something he or she has handled.

Animals   Flies carry germs onto food and eating utensils, whilst blood-sucking insects like mosquitoes inject germs under the skin.

Contaminated food   Sometimes germs get into food and water from sewage, or infected people. Typhoid is spread in this way.

Immunity to disease

You recover from a disease because your body produces substances called *antibodies*. These attack the germs or the poisons from the germs. Antibodies gradually disappear when their job is done, but your body is able to reproduce them very quickly if the germs invade again. You are then *immune* to the disease. It is now possible to produce immunity in other ways:

Vaccination   200 years ago Edward Jenner found that infecting people with the mild disease cowpox made them immune to smallpox, a fatal disease. Nowadays, a *vaccine* made from the skin of a calf infected with the vaccinia virus is placed on the arm. A scratch then made becomes infected and forms a spot which quickly heals as the antibodies build up. If a smallpox virus enters the body later, these antibodies will destroy it.

Inoculation   This is the injection of vaccine directly into the bloodstream. It may contain live germs made harmless by living in animals, dead germs, or extracts of the germ poisons.

Serums   These are made from the blood of horses which have become immune to a disease. They contain ready-made antibodies.

Jenner vaccinating his son against smallpox.

Sexually transmitted diseases

These are infections which can be passed from one person to another during sexual contact. Some of them are very serious because, if not treated, they cause severe damage to the body or even death.

Disease	Cause	Effects	Treatment
gonorrhoea (clap)	bacterium	pus/burning pain on passing urine – eventual sterility	cured by antibiotics
syphilis (pox)	bacterium	painless sores at first – eventual blindness, insanity and death	cured by antibiotics
herpes	virus	headache/temperature small painful blisters (cold sores)	keep clean avoid rubbing sores
AIDS	virus	tiredness, fever, weight loss, blotchy skin → death	none

AIDS (Acquired Immune Deficiency Syndrome)   is the most worrying of these diseases because there is no known cure. It is caused by the HTLV-III virus which stops the production of antibodies in some people. These people may die from illnesses which they would normally fight off easily. The virus is transmitted via blood or semen – during any form of sexual intercourse, when unscreened blood is transfused, or when drug abusers share needles.

Questions

1. What is meant by 'germs'? Draw a bacterium.
2. How do bacteria produce disease? Name three bacterial diseases.
3. How do viruses differ from bacteria? Name two virus diseases.
4. Describe four ways in which diseases are spread.
5. Explain how the body becomes immune to disease.
6. What is 'vaccination'? Who first developed the technique?
7. What is AIDS? Why is it so serious?

Useful bugs

Microbes (bacteria or fungi) are usually thought of as being harmful to people. Many of them are. Others are very helpful: they can produce foods, medicines, fuels, plastics and metals. The development of processes like these is called biotechnology.

Preserving food

If food is left about, it will be attacked by bacteria and fungi (page 202). Unpleasant or poisonous substances are formed. Most microbes will only flourish if they have air, water and gentle warmth. Food is preserved by denying the microbes the conditions they need – excluding air or removing water.

Killing germs	Taking air away	Taking water away
pasteurising – heating milk to 60 °C	**bottling** – baby foods, fruit	**drying** – fruit, milk, eggs, potatoes, packet soups
sterilising – boiling – adding sulphur dioxide or benzoic acid	**canning** – most foods	**freezing** – fish, meat, vegetables
	vacuum packing – coffee, peanuts	**salting** – bacon, meat
smoking	**carbon dioxide storage** – apples, vegetables	**sugaring** – candied peel fruit, jam
		vinegar pickling – onions

Making food

Certain microbes will change easily-decomposed foods into longer lasting forms. To do this, the microbes use *enzymes* in their cells. *Enzymes* are proteins which bring about chemical changes in substances around them whilst staying unchanged themselves. They are biological catalysts. There are thousands of enzymes and each one does a single specific job. Amylase, for example, is the enzyme in saliva which breaks down starch into easily absorbed glucose.

Yoghurt   When the enzyme in the bacterium *acidi lactiti* is added to milk it changes the sugar content of the milk into lactic acid. This makes the milk go thick and gives it the slightly sour yoghurt taste.

Cottage cheese   *Rennin* is an enzyme produced in the lining of the stomach and by certain fungi. It causes milk to separate into curds and whey (watery liquid). The curds set to form junket. This changes to cottage cheese when drained off and warmed. Milk always changes to junket in your stomach – you may have seen it if you have been sick soon after drinking milk.

Hard cheese   is made from soft cheese by drying, adding salt and squeezing it. Often a special kind of fungus is added to produce the particular taste of the cheese type.

Protein   is the vital part of our diet, obtained mainly from animal products. In Third World countries, protein is scarce and lack of it leads to the children's disease, Kwashiorkor. Rearing animals is costly and takes a long time. Now, however, protein can be made quickly and cheaply from potato or wheat starch using the mould *Fusarium*. Algae – the green slime in ponds – also provide a rich source of protein.

Microbes made this cheese

Yeast products

Alcohol   Yeast is a single-celled fungus which contains several useful enzymes. One of these, *zymase*, turns a solution of glucose into ethanol – usually called alcohol. The solution froths up because carbon dioxide is given off as the yeast grows. This is known as *fermentation*.

glucose $\longrightarrow$ ethanol + carbon dioxide

$$C_6H_{12}O_6 \xrightarrow[\text{water + warmth}]{\text{zymase}} 2C_2H_5OH + 2CO_2$$

All wines, beers and cider are made by fermenting fruits, vegetables or grain. Grapes contain a lot of natural glucose – other things are heated with water to change their starches and sugars into glucose. Fermentation stops when all the glucose has gone or when enough alcohol (about 12%) to kill the yeast has been produced.

Yeast like this helps to make bread, Marmite and alcohol.

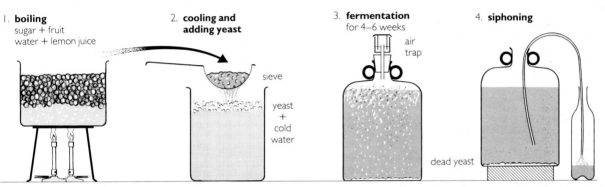

1. **boiling**
   sugar + fruit
   water + lemon juice

2. **cooling and adding yeast**
   sieve
   yeast + cold water

3. **fermentation** for 4–6 weeks
   air trap

4. **siphoning**
   dead yeast

Marmite   Yeast is a valuable source of protein and vitamin B. Much of the yeast produced in the brewing of beer is sold as animal food. Some is processed to form yeast tablets and extract (Marmite).

Bread   Dough is a mixture of flour, water and yeast. The yeast causes slight fermentation and the dough swells up with carbon dioxide bubbles. Baking kills the yeast and dries the dough.

Substitutes for oil

Today, most fuels, most plastics and many essential chemicals are obtained from crude oil. Alternatives will have to be found when the oil runs out. Already there are some processes which use microbes to make chemicals from easily grown plants (renewable resources).

Renewable resource	Substances made by microbes
sugar	hydrogen – fuel alcohol – fuel glycerol – fuel
starch	acetone – solvent butanol – plastics fructose – sweetener
cellulose	alcohol – petrol substitute methane – fuel

Questions

1. List eight different ways of preserving food.
2. What is an enzyme? What is amylase and what does it do?
3. Explain how cottage cheese is made.
4. State one way of making a meat-substitute protein.
5. What is yeast? How does it behave in a solution of glucose?
6. Explain the connection between Marmite and beer.
7. When making bread, what happens in the dough to make it rise?
8. Name five substances which microbes can make from renewable resources.

Healthy living

The human body is a very complex machine. Like all machines, it breaks down from time to time and parts wear out. If you want to live to be 100 you must treat your body with great care – eat sensibly, take exercise and avoid known health risks.

Diet

A balanced diet contains the five essential foods described on page 188 – carbohydrates, fats, proteins, vitamins and minerals. Young people can usually eat what they like because their bodies crave the foods needed for energy and growth. They like salty chips because these contain starch, fat and minerals. Eggs, sausages, baked beans and milk provide protein. Children sometimes run short of vitamins because they do not eat enough fruit or vegetables. These foods also contain fibre – lack of it leads to clogging up of the bowels!

As you get older, you need to control your diet. Too much of anything can be bad for you. Too much fat can cause artery and heart problems. High fibre and low fat diets are most healthy.

Food additives

These are substances added to food to preserve it, to make it look nicer or to taste better. Some additives occur naturally and others are made artificially. Sugar and salt have been used for thousands of years to stop food going bad and to enhance its flavour. Today, all sorts of things are used, particularly in made-up (junk) foods. Packet foods must be clearly labelled with everything in them. The additives have all been tested and those thought to be safe have been given E numbers. Those still being tested have numbers but no E in front.

In spite of the tests, some people are affected badly by certain additives. The yellow colour, tartrazine (E102) is thought to cause skin problems, breathing difficulties and hyperactivity in some children.

A healthy diet has plenty of fibre and not much fat.

BEST BEFORE: 10 JAN 87

CRISPLY FRIED BACON FLAVOUR CORN SNACK

Ingredients: Maize; Rice Flour; Vegetable Oil with antioxidants E320, E321; Seasoning (flavour enhancer monosodium glutamate, flavouring); Hydrolysed Vegetable Protein; Salt; Sugar; Magnesium Carbonate; Trisodium Phosphate; Colours E110, E127,133.

Type of additive	E numbers	Example of type	
		Natural	**Possibly harmful**
Colourings	100–180	E 162 beetroot red	E 102 tartrazine
Preservatives	200–297	E 270 lactic acid	E 210 benzoic acid
Anti-oxidants	300–321	E 300 vitamin C	E 311 octyl gallate
Emulsifiers and stabilisers	322–495	E 330 citric acid	E 450 (c) sodium polyphosphate
Gelling agents and others	500–578 900–925	516 calcium sulphate	544 calcium polyphosphate
Flavourings	620–637	620 L-glutamic acid	621 monosodium glutamate

Health risks

Smoking is the most serious health risk in the UK. Each year, 100 000 people die because they smoke 20 or more cigarettes a day. These heavy smokers are likely to die from heart attacks, blocked arteries, lung diseases or cancer.

Not all smokers will die from these causes but all will suffer from bad coughs, raised blood pressure and less oxygen in the blood. The wind-pipe and lungs are kept clean by *cilia* – tiny hairs covered with mucus. These trap germs and dirt and move them up to the mouth as spit. Smoking paralyses and eventually destroys the cilia. The lungs clog up.

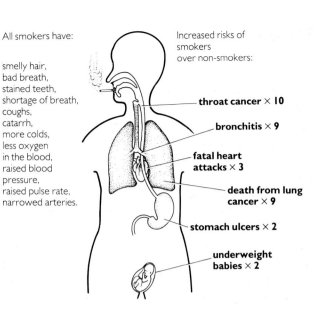

All smokers have:

smelly hair,
bad breath,
stained teeth,
shortage of breath,
coughs,
catarrh,
more colds,
less oxygen
in the blood,
raised blood
pressure,
raised pulse rate,
narrowed arteries.

Increased risks of smokers over non-smokers:

throat cancer × 10

bronchitis × 9

fatal heart attacks × 3

death from lung cancer × 9

stomach ulcers × 2

underweight babies × 2

Drugs are used by doctors to treat diseases and mental depression or to relieve pain. When taken in large doses by people who do not need them, the same drugs can cause severe health problems, personality changes or death. Drugs make people feel excited or relaxed – some cause a dreamy trance. In these states, people lose control of themselves and can have serious accidents. When the effects wear off, there are often bad mental or physical reactions. Drug takers need more and more drugs and can become totally dependent upon them – they become addicts (junkies). Many addicts are young and die before they reach 30.

Glue sniffing is very dangerous. The vapours act like drugs and also do great damage to the lungs and brain. Young people have died after sniffing quite small amounts of solvents.

Alcohol is a depressant drug. One small glass of beer, whisky or sherry relaxes people and makes them more sociable. After two or three drinks, control is lost over speech and movement. It is dangerous to drive. The legal limit is 80 mg alcohol in 100 cm^3 blood. Large quantities of alcohol will cause coma or even death. Years of heavy drinking can damage the stomach, liver and heart.

Mis-used drugs
1. **Depressants** – induce relaxation and reduce worry. Examples are: heroin (junk) barbiturates (barbs) tranquilizers (sleepers)
2. **Stimulants** – give feelings of energy and confidence. Examples are: cocaine (coke) amphetamines (purple hearts)
3. **Hallucinogens** – increase self awareness and reaction to colour or sound. Examples are: cannabis (grass, pot, marijuana) L.S.D. (acid) psylocybin (magic mushrooms)

Questions

1. State two reasons for eating plenty of fruit and vegetables.
2. What sort of diet reduces the chances of heart disease?
3. Give three reasons for using food additives. What effects may E102 have?
4. Why do smokers cough? What illnesses are they likely to die from?
5. What is a drug addict? Name two dangerous drugs.
6. What problems may be caused by drinking heavily for a long time?

Further questions

1 Give three important differences between living and non-living things.

2 A plant cannot move about to search for food. How does a plant obtain its food?

3 Why do plants and animals need oxygen?

4 **a** Copy the diagram of the plant cell and complete the labels.
b State two labels which should not be included in a diagram of an animal cell. <small>EAEB</small>

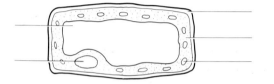

5 Copy and complete the word equation showing what happens in the leaf of a plant during photosynthesis:

$$\ldots + \text{water} \xrightarrow[\text{from}]{\text{using energy}} \text{food} + \ldots$$

6 What green material in plants enables photosynthesis to take place?

7 What is a transpiration stream? Why is it important?

8 Why does the amount of starch in the leaf of a plant not remain constant?

9 Give one function of:
a the root of a plant; **b** the stem; **c** the leaves. <small>WYLREB</small>

10 Which of the following plants do not have flowers?
grass, fern, daisy, oak, moss.

11 Which of the following was last to appear on Earth?
reptile, fish, mammal, bird, amphibian. <small>WMEB</small>

12 **a** On this diagram the labels have got muddled up. Redraw the diagram and put them in the right places.

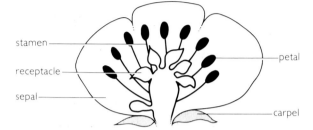

b In which part of the flower is pollen produced?
c Why is nectar produced by the flower?
d What happens in a flower during pollination? In what ways can pollination occur?
e What process follows after pollination has taken place?

13 Why do plants scatter their seeds over a large area?

14 For a seed to germinate, which of the following are needed and which are not?
moisture, light, warmth, air. <small>WY & LREB</small>

15 Why does an indoor potted plant turn its leaves towards the window?

16 State two ways in which nitrogen in the atmosphere is made available to plants for food making. <small>ALSEB</small>

17 What food substances are made by plants using the nitrogen they take in?

18 In the human body, how are the following protected from outside damage?
a The heart and lungs **b** the brain **c** the spinal cord

19 What substance give bones their strength?

20 Where in a tooth are the *pulp, enamel* and *dentine*?

21 Name the four types of teeth present in the jaws of an adult human. What is the main purpose of each type?

22 How is a knee joint different from a hip joint?

23 What are antagonistic muscles? Give an example.

24 Give an example of **a** a voluntary muscle movement **b** an involuntary or reflex muscle movement.

25 **a** What are the five main types of food or nutrient?
b What is meant by a balanced diet?

26 Describe two features of the small intestine which help it to absorb digested food.
How is digested food transported to the cells of the body?

27 What is the difference between blood and blood plasma?

28 What jobs are done by **a** red blood cells **b** white blood cells?

29 **a** Why can the heart be described as 'two pumps in one'?
b What is the difference between a vein and an artery?
c Which organ receives blood through the pulmonary artery?

30 Copy and complete the following:
The lungs hang in an airtight cavity called the
.... When you breathe in, this cavity is expanded in two ways – the rib cage is pulled upwards and outwards by .... muscles, and a curved sheet of muscle called the ....
pulls downwards. As you breathe in and out, ... gas is absorbed into the blood and .... gas is released into the air.

31 **a** Copy the outline of the drawing of the human body.

Draw in and label the following organs: *heart*, *lungs*, *liver* and *kidneys*   ALSEB

b Which of the above are organs of excretion?

32 **a** What is meant by **i** fertilisation **ii** the menstrual cycle?
b The diagram shows part of the female reproductive system.

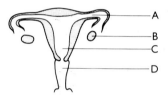

Name the parts labelled A to D.
In which part does fertilization take place?
In which part does the foetus develop?   ALSEB

33 The diagram below shows some energy exchanges. What words go in the boxes A, B, C, D and E? Choose from this list:

respiration   *combustion*   *photosynthesis*
eating   *decay*

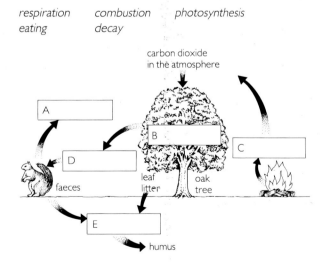

34 a One way of sorting animals is to group them according to the way they move. Six animals are shown in the diagram.

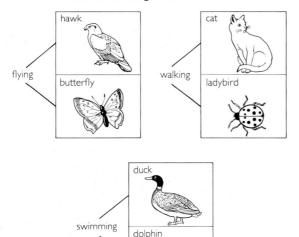

Copy and complete the table below by rearranging these animals into three different groups. Give *one* reason in each case for placing each animal in these new groups. You need not draw the animals.

Name of group	Name of animal	Reason for the animals being in this group
birds	hawk	both have feathers/beak lay eggs
	duck	

b Hawks eat smaller birds which in turn eat animals such as ladybirds. Ladybirds eat greenfly. When people use insecticides to protect crops against greenfly, how can this harm hawks? Explain your answer. LEAG

35 Many of the foods in supermarkets will be 'preserved' in some way. There are many methods available to food producers but the method must suit the food. The following table gives the methods used for some popular foods.

	Dried	Pickled	Pasteurised	Chemical preservative	Smoked	Sterilised and sealed	Inert gas
Corn flakes							
Baked beans							
Mackerel							
Sausage							
Onions							
Crisps							
Milk							

a Copy the table and complete it by placing a tick in the correct column for each of the foods not yet marked.

b i It is likely that if manufacturers no longer added preservatives the price of foods would *not* come down. Explain this.   **ii** How do food preserving methods make life easier for a supermarket manager? LEAG

36 *penis, vagina, thirty, placenta, sperm, ovum, foetus, embryo, uterus, fallopian, nine, forty*

Choose the most suitable word in the list to complete the passage below.

In the process of fertilisation the male cell called the . . . . fuses with the female egg called the . . . . This forms a cell which divides into a ball of cells called the . . . . which later develops into the . . . . After about . . . . months of pregnancy, signs of birth become evident and contractions of the . . . . muscles indicate the start of labour. MEG

37 Below is an example of a food chain:

grass → cattle → humans

In an area of saltmarsh, wild ducks were seen feeding on small snails. Foxes raid the nest of the ducks for chicks and eggs. Small snails feed on the green algae which cover the surface of the mud.

a Use the information provided to show the food chain for the saltmarsh.

b In a similar area of saltmarsh, hunters shot many of the ducks. How might this affect:
i the foxes   **ii** the small snails   **iii** the algae? L

Chapter 6

Currents and magnets

In one way it makes your hair stand on end, in another it dries it. What is it?

The light for this photograph came from two sources . . .

Electric charge

It passes through the wires when you switch on the light and it causes lightning flashes. It makes polythene film stick to your hands and dust stick to your records. It can even make your hair stand on end!

It's called electric charge – no one knows what it is but a great deal is known about where it comes from and the way it behaves.

The electric atom

In the centre of an atom there is a *nucleus* made up of particles called *protons* and *neutrons*. Around this nucleus orbit very much lighter particles called *electrons*.

Electric charge is carried by the electrons and protons in every atom. There are different types of charge:

the charge on an electron is called a *negative* (−) charge;
the charge on a proton is called a *positive* (+) charge;
There is no charge on a neutron.

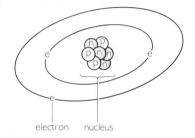

electron   nucleus

Normally, atoms have equal numbers of electrons and protons – they have equal amounts of − and + charge within them.

Electrons do not always stay attached to atoms. When you switch on a light, the 'electricity' that travels through the wires is actually a flow of electrons.

Conductors and insulators

Materials that allow electrons to pass through them are called *conductors*. All metals are good conductors of electricity, so is carbon. In a conductor, some of the electrons are not very tightly held to the atoms and are free to travel through the material.

Most non-metals conduct poorly or not at all. Materials that do not normally allow electrons to pass through them are called *insulators*: their electrons are all tightly held to atoms and are not free to move.

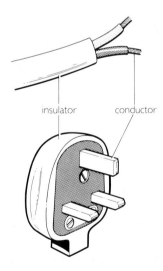

insulator   conductor

Conductors		Insulators	
Good	*Bad*	rubber	glass
metals	water	plastics	dry air
especially	the human body	e.g.	
silver	earth	PVC	
copper		polythene	
aluminium		Perspex	
carbon			

Static electricity

If a material gains or loses electrons, there is no longer an exact balance between the − and + charges within it – the material is said to be *charged*. Charged materials are sometimes said to have 'static electricity' on them.

polythene

Charging by rubbing Insulators become charged when they are rubbed:

A piece of polythene gains − charge when rubbed with a cloth. The polythene pulls electrons away from atoms in the cloth, leaving itself with more electrons than normal and the cloth with less. The charge stays on the polythene because polythene is an insulator – the extra electrons are unable to flow away through the material, though in time they leak into the air.

cellulose acetate

A sheet of Perspex is left with a + charge when rubbed with a cloth. The cloth pulls electrons away from some of the atoms in the sheet, leaving the atoms with more protons than electrons.

Forces between charges An electric charge will push or pull on any other charge nearby.

If two charged strips of polythene are held together at one end, the strips are pushed apart:

Like charges repel each other If you pull a polythene bag quickly through your hand, the polythene gains a − charge and your hand is left with a + charge. The bag clings to your hand because the − and + charges pull on each other:

Unlike charges attract each other The attraction between unlike charges makes dust cling to records and can even cause sparks when you take off nylon or Terylene clothing – the sparks occur when electrons are pulled strongly enough to make them jump through air.

repulsion

attraction

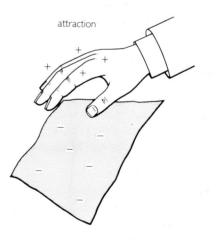

Questions

1. What type of charge is on   (a) an electron;   (b) a proton?
2. A sodium atom has 11 protons. How many electrons does it have?
3. What actually travels through the wires when you switch on a light?
4. What is the difference between a conductor and an insulator? Name two good conductors.
5. What insulator would you rub with a cloth to produce   (a) a negative charge;   (b) a positive charge?
6. How could you show that like charges repel each other?
7. How could you show that unlike charges attract each other?

Electric cells

Electric cells, often called batteries, are a very convenient source of electricity and can be used to power many things from radios to submarines. The electrons they release come from chemical reactions that take place inside them.

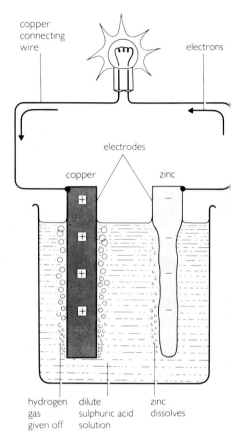

copper connecting wire

electrons

electrodes

copper   zinc

hydrogen gas given off   dilute sulphuric acid solution   zinc dissolves

The simple cell

A simple cell can be made by placing a zinc plate and a copper plate in a dilute solution of sulphuric acid. The cell will light up a small bulb if there is a conducting path from one plate to the other plate through the bulb. As copper is a good conductor, copper wires can be used to connect the bulb to the plates.

The zinc plate dissolves in the acid, releasing electrons as it does so. These electrons leave the zinc plate and pass through the bulb, making it glow brightly. On reaching the copper plate, the electrons cause hydrogen from the acid to form into hydrogen gas which bubbles off the plate. Overall, the zinc goes into solution, the hydrogen comes out of solution, and electricity is released in the process.

The two plates are known as *electrodes*:

the zinc plate is the *negative (−) electrode* – it acts as a store of electrons;
the copper plate is the *positive (+) electrode* – it is always ready to receive electrons.

Polarisation   The bulb does not glow brightly for long. A blanket of bubbles builds up round the copper plate and reduces the flow of electrons from the cell – an effect known as *polarisation*. A *depolarising agent*, such as potassium dichromate, can be added to the solution to remove hydrogen from the copper plate; when this is done, the bulb begins to glow brightly again.

The dry cell

The batteries you find in the handle of a torch are of a type known as *dry cells*. Instead of an acid, a dry cell uses ammonium chloride in 'dry' jelly form, so the cell does not have to be kept upright.

The zinc case of the cell acts as the negative (−) electrode; the carbon rod in the centre of the cell is the positive (+) electrode. Around the carbon rod is the depolarising agent – a mixture of manganese dioxide and powdered carbon.

The simple cell and the dry cell are both called *primary* cells – they will release charge but they cannot be recharged. *Secondary* cells on the other hand store charge that is put into them – they can be recharged over and over again.

carbon rod

brass cap

ammonium chloride jelly

manganese dioxide/ carbon

zinc case

Lead-acid storage cells

The lead-acid cell, or *accumulator*, is a secondary cell. It has two sets of plates, each connected to a *terminal*. Both sets are made of lead but the positive plates are coated with lead dioxide paste. The plates are in a dilute solution of sulphuric acid.

The cell discharges (gives out electrons) as the lead and the lead dioxide on the plates are slowly converted to soft lead sulphate. The reaction dilutes the acid making it dense.

When the cell is connected to a battery charger, electrons are pushed back through the cell. This reverses the chemical reactions – lead, and lead dioxide, are built up on the plates again.

Batteries Single cells are often called batteries, though a *battery* really means a collection of cells. A car battery, for example, is made up of six lead-acid cells connected one after another in a row. Together, the cells push out electrons with greater force than a single cell.

Care of car batteries Nowadays, most batteries are 'low maintenance' and need very little attention. However, it is important not to leave a car battery 'flat' for any length of time. A 'flat' (fully discharged) battery soon becomes 'sulphated' and cannot then be recharged. A 'sulphated' battery is one in which the soft lead sulphate on the plates has changed into a hard form that cannot be converted back into lead and lead dioxide. With some batteries, a garage can check the state of charge of each cell of a battery by measuring the relative density of the acid solution with a hydrometer (see page 143); as a cell loses its charge, the density of the acid gets less.

In a car, an alternator turned by the engine keeps the battery charged up. If a battery is fully charged, any further charging turns some of the water into hydrogen and oxygen gas bubbles. The car charging system should shut off before this happens. However, if any water is lost, the battery must be topped up with distilled water to replace it.

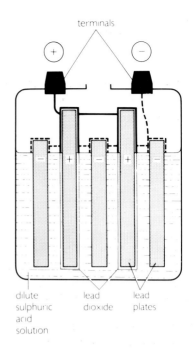

dilute sulphuric acid solution / lead dioxide / lead plates

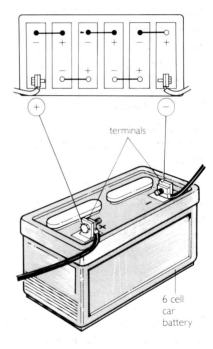

terminals

6 cell car battery

Questions

1. In a simple cell, what materials are used for the + plate and the − plate? In what liquid are they placed?
2. What is polarisation? How does it affect a simple cell? How can polarisation be prevented in a simple cell?
3. What are the positive and negative electrodes of a dry cell made of? What depolarising agent is used in the cell?
4. How is a secondary cell different from a primary cell?
5. How could a garage check the state of charge of a car battery? Why must the battery not be left 'flat' for a long time? Why must distilled water sometimes be added to a car battery?

Current and voltage in a simple circuit

Connect a bulb to a battery and electrons flow through the bulb. But what have amps and volts got to do with electron flow?

A simple circuit

The diagram on the right shows a light bulb connected by two conducting wires to a battery. The bulb and the battery have been drawn using symbols – the symbol for a single cell is also shown.

The conducting path through the bulb, wires and battery is called a *circuit*. Electrons flow around this circuit from the − to the + terminal of the battery.

In this book, an electron flow is shown by a series of arrows drawn outside the circuit itself.

Current

Electrons flowing round the circuit each carry a tiny − charge. The flow of charge is called a *current* and it can be measured by placing an instrument called an *ammeter* in the circuit.

Current is measured in *amperes* (A), or 'amps' for short. If a current of 1 ampere (1 A) flows in a circuit, about 6 million million million electrons pass round the circuit every second.

The current through a small torch bulb is about ¼ A; the current through a car headlight bulb is about 4 A.

Small currents can be measured in *milliamperes* (mA):

1000 mA = 1 A

Current values round a circuit   In the diagram on the right, an extra bulb has been added to the circuit, together with two more ammeters.

All three ammeters show the same reading. Electrons leaving the battery all pass through every section of the circuit, so the same current is measured at all points.

Current direction   Many people regard current as a flow of + charge from the + to the − terminal of a battery – a convention decided after the first batteries had been made and used but before electrons had been discovered.

In this book, the direction of the conventional current is shown by a series of arrowheads drawn on the circuit itself. You can see from this diagram that the electrons flow in the opposite direction to the conventional current.

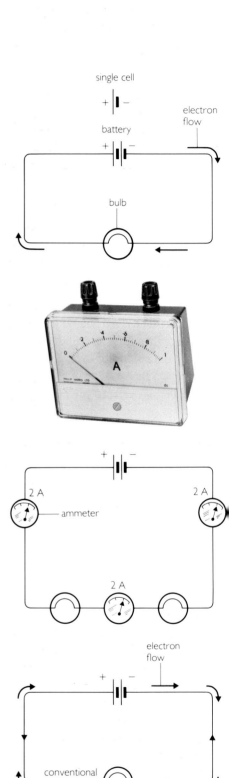

Energy from a battery

In the circuit on the right, the battery acts as an electron 'pump': it pushes electrons out of the − terminal, giving them energy which they carry to the bulb. The electrons lose all this energy as they are forced through the tiny coiled wire inside the bulb. The energy is radiated in the form of light and heat.

Voltage

Battery voltage   Some batteries push out electrons with greater force than others, giving them greater amounts of energy in the process. The higher the *voltage* across a battery, the more energy is given to each electron pushed out. The voltage can be measured by connecting an instrument called a *voltmeter* across the battery terminals.

Voltage, or potential difference (p.d.), is measured in volts (V)

The voltage across a dry cell is about 1½ volts (1½ V); the voltage across a car battery is about 12 volts (12 V). Electrons pushed out from a car battery have more energy than electrons pushed out from a dry cell.

Voltages around a circuit   The circuit on the right includes two bulbs. Electrons lose part of the energy given them by the battery, in passing through the first bulb. They lose the rest of it in passing through the second bulb.

Connect a voltmeter across either bulb and it will show a reading. The higher voltage across the bulb, the more energy each electron loses as it pushes through it.

Between them, the two bulbs use all the energy supplied by the battery:

the voltages across the two bulbs add up to equal the voltage across the battery.

electron flow

electrons given energy

electrons lose energy

4 V

+ −

2 V   2 V

Questions

1. What is a flow of charge called?
2. (a) Draw the circuit on the right, adding instruments to measure the current through the bulb and the voltage across the battery. What is each instrument called?
    (b) Where in the circuit do the electrons lose their energy?
    (c) If the current through the bulb is 2 A, what is the current through the battery?
    (d) If the voltage across the battery is 4 V, what is the voltage across the bulb?
3. What is each of the following in A?
    1000 mA   500 mA   2500 mA

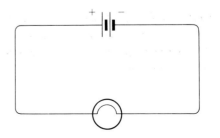

+ −

Resistance

Electrons pass easily through copper connecting wire. They don't pass so easily through the tiny coiled wire in a light bulb or the wire used in an electric fire element. These wires have much more resistance.

Resistance

The wire used in electric fire elements is made of a nickel chromium alloy called nichrome. Energy is needed to push electrons through nichrome wire; the wire is said to have *resistance*.

A long piece has more resistance than a short piece;
a thin piece has more resistance than a thick piece;
a hot piece has more resistance than a cold piece.
A piece of nichrome wire has much more resistance than a same size piece of copper wire – nichrome is a poorer conductor than copper.

Resistors   These are devices specially made to provide resistance; placed in a circuit, they reduce the flow of current. A length of thin nichrome wire makes a simple resistor; other types of resistor are shown on the right.

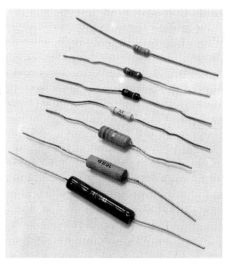

Ohms

Resistance is measured in *ohms*. A resistor has a resistance of one ohm ($1\ \Omega$) if a voltage of one volt across it will push a current of one ampere through it.

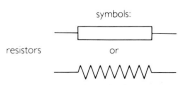

symbols:

resistors   or

The more resistance a resistor has, the more volts are needed to push each ampere through. If the voltage across a resistor and the current through it have both been measured, the resistance (R ohms) can be calculated using the equation:

$$R = \frac{V}{I}$$

where V is voltage (volts)
and I is current (amperes)

On the right, the equation has been asked to calculate the resistance of the resistor shown in the circuit.

Resistor calculations   The equation can be rearranged to give:

$$V = I \times R \quad \text{and} \quad I = \frac{V}{R}$$

These equations can be used to calculate a voltage or current if the value of the resistance is already known. The triangle on the right gives all three equations – cover each letter in turn with your thumb.

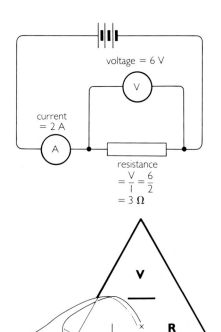

voltage = 6 V

current = 2 A

resistance
$$= \frac{V}{I} = \frac{6}{2}$$
$$= 3\ \Omega$$

Ohm's Law   Increase the voltage across a resistor and more current flows. In the 1820's, George Ohm discovered that the current through a metal resistor increased in direct proportion to the voltage across it provided the metal was kept at a fixed temperature.

In other words, if 6 volts produced a current of 2 amperes through the metal, 12 volts would produce a current of 4 amperes – the resistance of the metal would remain unchanged at 3 ohms provided the metal was kept at the same temperature.

Heat and resistance

Any conductor which has resistance gives off heat when a current flows through it. As electrons are pushed through the conductor, they collide with the atoms making them vibrate more vigorously.

Heating elements   Nichrome wire, coiled to take up less space, is used to make the heating elements found in many electrical appliances used in the house. Not only does nichrome provide high resistance, it can be kept red hot without oxidising and breaking in the air.

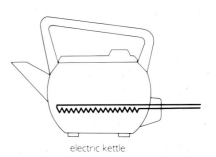

electric kettle

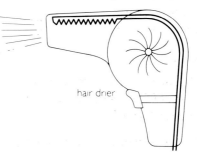

hair drier

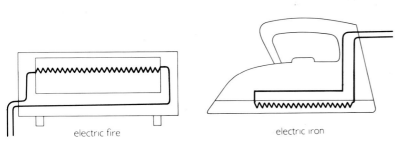

electric fire   electric iron

Light bulbs   The *filament* of a light bulb is made from very thin tungsten resistance wire. When a current is passed through it, the tungsten becomes so hot that it gives off a brilliant white light. Having a very high melting point, the tungsten can be kept white hot without melting. In air, the hot filament would quickly burn up, so the bulb is filled with argon and nitrogen gases which do not react with the hot metal.

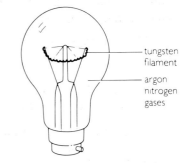

tungsten filament

argon nitrogen gases

Questions

1. Which has more resistance   (a) a long piece of nichrome wire or a short piece?   (b) a thick piece or a thin piece?
2. What do the symbols V, I and R stand for? What equations link them?
3. A metal resistor has a voltage of 12 V across it and a current of 2 A through it. What is its resistance? If the voltage drops to 6 V what current would then flow?
4. What voltage is needed to push 3 A through an 8 Ω resistor?
5. What happens when a current flows through a resistance?
6. What is the filament of a bulb made of? Why is this material used? Why is a bulb not filled with air? What does it contain?

Circuits and switches

How can you run two bulbs from one battery? Various different methods of connecting bulbs, switches, resistors and batteries have very different effects.

Series and parallel circuits

Connected to a battery, a single bulb glows brightly. There are two ways in which a second bulb can be added to the circuit:

Bulbs in series   In the left-hand diagram, the two bulbs have been connected in *series*.

Together, the two bulbs provide greater resistance than a single bulb and less current flows as a result. Both bulbs glow dimly – the available voltage has been 'split' between them.

The right-hand diagram shows what happens when one of the bulbs is removed. The circuit is broken, so the other bulb goes out.

Bulbs in parallel   In the left-hand diagram, the two bulbs have been connected in *parallel*.

Both bulbs glow brightly because each has the full voltage of the battery across it. Together, the two bulbs take twice as much current as a single bulb, so the battery loses its charge twice as quickly.

The right-hand diagram shows the effect of removing one of the bulbs. There is still an unbroken circuit through the other bulb, so it continues to glow brightly.

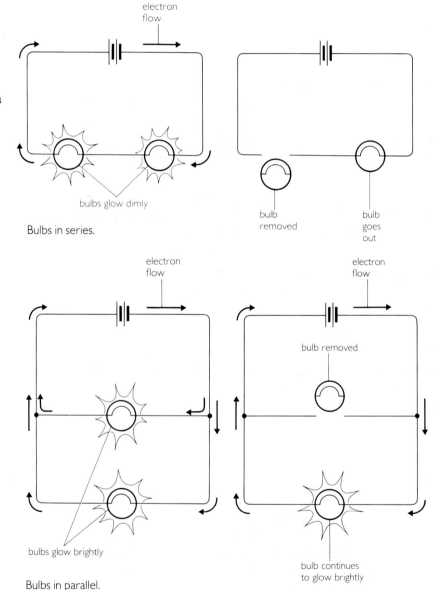

electron flow

bulbs glow dimly

Bulbs in series.

bulb removed

bulb goes out

electron flow

electron flow

bulb removed

bulbs glow brightly

bulb continues to glow brightly

Bulbs in parallel.

Bulbs, electric fires and other appliances used in the house are always connected in parallel with the mains supply. Each then receives the full mains voltage and each can be switched on or off without affecting the others.

Switches

A switch breaks a circuit by moving contacts apart. The diagram shows how switches might be placed in a circuit to control a number of different light bulbs.

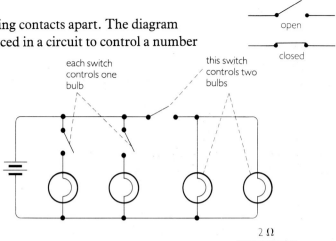

To check which switch controls each bulb, trace a route with your finger from one side of the battery through a bulb to the other side of the battery. The bulb is switched on and off by whichever switch your finger passes over.

each switch controls one bulb

this switch controls two bulbs

Resistors in series and parallel

In series, two resistors have a combined resistance found by adding up their resistance values. Using two resistors instead of one has the same effect as using a long piece of nichrome wire instead of a short piece – it gives the electrons a more difficult conducting path to travel through.

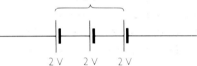

If two identical resistors are connected in parallel, their combined resistance is only half that of a single resistor. The effect is the same as using a thick piece of nichrome wire instead of a thin piece – the electrons are given a wider path to travel through.

Cells in series and parallel

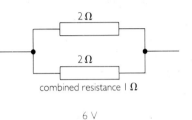

If three 2 volt cells are connected in series, the voltage across all three is 6 volts. Together, the three cells give electrons more energy than a single cell.

When the three 2 volt cells are connected in parallel, the voltage across them is only 2 volts. The cells will last longer than a single cell however, and are together capable of supplying a higher current.

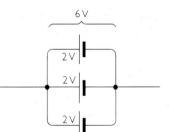

Questions

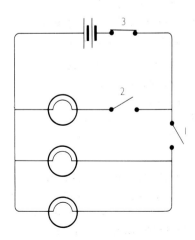

1. If you were going to connect two light bulbs to one battery, would you use a series or a parallel arrangement? Why? Which arrangement takes most current from the battery?
2. In the circuit on the right, what happens to each of the bulbs when you   (a) close switch 1;   (b) then close switch 2;   (c) then open switch 3?
3. Show how you would connect two 4 ohm resistors to produce a combined resistance of   (a) 8 ohms;   (b) 2 ohms.
4. Show how you would connect two 1 volt cells to produce a voltage of   (a) 2 volts;   (b) 1 volt.

221

Transistors

Transistors are used in TV's, record players, and just about every other piece of electronic equipment you can think of. They can link circuits together in such a way that the current through one controls the current through the next.

You can see some transistors in the photograph. Inside each case is a specially treated crystal or *chip* of silicon, with three connecting leads attached.

Amplifying and switching

In a record player, transistors are used to magnify or *amplify* the tiny, changing current generated by the pick-up. The changing current in one circuit causes even bigger changes in the next and so on . . . This continues until the changes are large enough to make a loudspeaker cone vibrate.

Transistors are also used as switches. For example, in many washing machines they switch on and off the circuits which control the wash program. You can find out more about transistor switches in the next unit.

Transistors are very small.

Integrated circuit (IC): thousands of transistor circuits on a single chip of silicon.

ICs are used in this stereo system.

The photograph above shows an *integrated circuit* (IC). This contains many complete circuits, with transistors, resistors, connections, and other components all formed on a tiny chip of silicon only a few millimetres square. Most of the transistors in a stereo system or washing machine are in *microchips* like this.

A switched off transistor . . .

The transistor in the circuit on the right is an *npn junction* transistor. It is shown using a symbol. It has three terminals. These are the *emitter*, the *base*, and the *collector*.

Connected like this, the transistor won't conduct. So the bulb doesn't light. The transistor is 'switched off'.

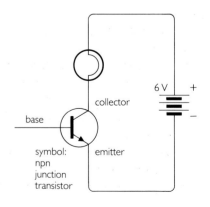

Switching the transistor on

Put a small voltage across the base and emitter of the transistor, and a tiny current flows through the base. This current completely alters the way the transistor behaves. The transistor starts to conduct, and the bulb lights up. It takes a base current of less than 1 mA to 'switch on' the transistor so that the bulb lights.

The transistor won't switch on unless the + and − connections are the right way round. The *conventional* current direction must be the *same* as the *arrow* direction in the symbol.

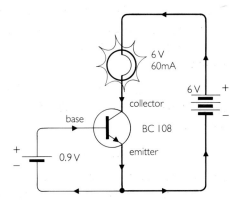

Another way of switching it on

You don't need a second battery to switch a transistor on. With a *voltage divider*, you can use part of the main battery voltage instead. This is happening in the circuit on the right.

The voltage divider is made up of two resistors in series. Together, these have the full 6 volts from the battery across them. The lower resistor is **variable** – by turning a knob, its resistance can be changed from zero up to 10 000 ohms.

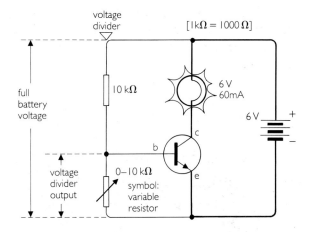

If the variable resistor is set to zero, it gets no share of the battery voltage. The output from the voltage divider is zero, so the transistor stays switched off.

If the variable resistor is turned 'up', it gets a bigger share of the battery voltage. So the output from the voltage divider rises. When it passes 0.6 volts, the transistor switches on and the bulb lights.

Questions

1. In the circuit on the right, the bulb is glowing brightly.
   a Which letter, A to E, stands for each of these?
   base   *collector*   *battery +*   *emitter*
   b Draw the circuit diagram. Mark on arrows to show the conventional current direction in each part of the circuit.
   c What do the components G and H do?
   d What would happen if you replaced H with a short piece of connecting wire? Explain why.

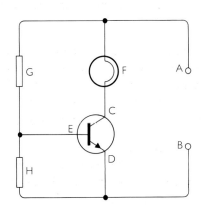

Transistor switches

To turn a circuit on, you don't always have to press a switch. With a transistor, it can be turned on by heat or light instead.

In the shop in the photograph:
> the doors open when a customer approaches
> the entrance lights come on when darkness falls
> if there is a fire, an alarm bell sounds.

These all happen automatically. They're all controlled by transistor switches. The circuits below show the basic principles.

A light-operated switch

This circuit uses a *light-dependent resistor* (LDR). This is a special type of resistor whose resistance falls when light shines on it.

With this circuit, the bulb lights up when the LDR is put in the dark.

The LDR is part of a voltage divider. In daylight, the LDR has a low resistance, and a low share of the battery voltage. The output from the voltage divider isn't enough to switch the transistor on.

In darkness, the resistance of the LDR rises. This gives the LDR a much larger share of the battery voltage. Now, the output from the voltage divider is more than the 0.6 V needed to switch the transistor on. So the bulb lights up.

If you replace the top resistor with a variable resistor, you can alter the 'darkness level' needed to switch the transistor on.

If you swap round the LDR and the top resistor, the bulb will be 'off' in the dark and 'on' in the light.

Automatic doors.

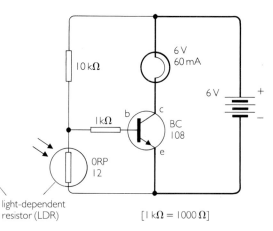

light-dependent resistor (LDR)

10 kΩ

1 kΩ

ORP 12

6 V 60 mA

BC 108

6 V

$[1\,k\Omega = 1000\,\Omega]$

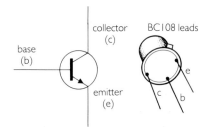

base (b)

collector (c)

emitter (e)

BC108 leads

A heat-operated switch

This circuit uses a *thermistor*. This is a special type of resistor whose resistance fails when it is heated.

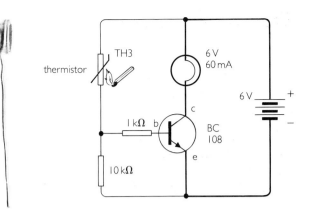

With this circuit, the bulb lights up when you heat the thermistor with a match.

The thermistor is part of a voltage divider. At room temperature, the thermistor has a high resistance. The output from the voltage divider isn't enough to switch the transistor on.

When the thermistor is heated, its resistance falls. This gives the bottom resistor a larger share of the battery voltage. Now, the output from the voltage divider is more than the 0.6 V needed to switch the transistor on. So the bulb lights up.

If you replace the bottom resistor with a variable resistor, you alter the temperature needed to make the transistor switch on.

Using a relay

The transistor in the circuit above would be damaged by a large current. So it can't be connected directly to an alarm bell or a set of shop lights. These have to be connected using a *relay* like the one in the photograph. The transistor switches on the relay and the relay switches on the other circuits. You can find out more about relays on page 237.

Relay.

Questions

1. The chart on the right gives information about four transistor switch circuits. It tells you what is needed for the transistor to be switched ON or OFF.

   Which circuit is most like the one that would be used to:

   a open an automatic door when a light beam is cut?

   b set off an alarm when a fire starts?

   c switch on a car parking light at dusk?

   d switch on a heater when the temperature in a greenhouse falls?

	transistor **ON**	transistor **OFF**
circuit A	darkness	bright light
circuit B	bright light	darkness
circuit C	high temperature	low temperature
circuit D	low temperature	high temperature

Mains electricity

When you plug in an electric kettle, you are connecting it into a circuit. The circuit hasn't got a battery in it but the mains supply is doing much the same sort of job.

A simple mains circuit

The diagram shows the circuit formed when a kettle is plugged into a mains socket. All the wires to the kettle are insulated and contained in a single cable or 'flex':

The alternating mains   The current from the mains socket is not a one-way flow like the current from a battery; it is pushed and pulled forwards and backwards through the circuit 50 times every second. The current is known as *alternating current* (AC). Power stations supply AC because it is easier to generate than one-way *direct current* (DC).

AC is supplied to the mains socket at a voltage of 240 volts.

The live wire   This goes alternatively − and + as electrons are pushed and pulled through the circuit.

The neutral wire   The Electricity Board *earth* the neutral wire by connecting it to a metal plate buried in the ground. Although current passes through the wire, any charge that might build up on it drains away into the ground – if you accidentally touch the neutral wire, you don't get a shock.

The earth wire   The metal body of the kettle is connected to a third wire called an earth wire – in many houses, the wire is earthed by connecting it to the neutral wire in the supply cable.

The earth wire is a safety wire which prevents the kettle becoming 'live' if a fault develops – current flows to earth rather than through anyone who happens to touch the kettle.

The switch   The switch on the mains socket is fitted in the live wire. If it were in the neutral, wire in the flex to the kettle would be 'live' even with the switch turned off.

The plug   This provides a convenient and safe way to connect up different appliances. There are eight stages in wiring up a plug. The first four are shown on the right, and the second four in the next section.

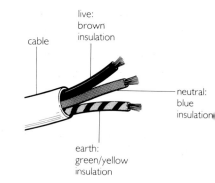

The fuse This is a short piece of thin wire that overheats and melts if too much current flows through it. Like the switch it is placed in the live wire, often in the form of a small cartridge inside the plug. If too high a current flows because of a fault in the kettle, the fuse 'blows' and breaks the circuit before the cable can overheat and catch fire.

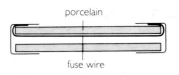

porcelain

fuse wire

Power

The electric kettle changes electrical energy from the mains socket into heat energy. The amount of energy it changes every second is called the *power* and it is measured in *watts* (W).

A power of one watt (1 W) means that the energy is being changed at the rate of one joule every second.

If the kettle has a power of 2400 watts, it is turning 2400 joules of electrical energy into heat every second. The chart on the right gives the powers of several common household appliances.

Power can also be measured in kilowatts (kW):

1 kilowatt = 1000 watts.

The power equation The power of the electric kettle depends on the voltage across the kettle and the current through it: the greater the voltage, the more energy each electron brings; the greater the current, the more of these energy-bringing electrons pass through the kettle every second.

The power of any appliance is given by the equation:

power = **voltage** × **current**
(watts)   (volts)   (amperes)

The equation shows, for example, that a 2400 watt kettle will draw a current of 10 amperes from the 240 volt mains:

2400 = 240 × 10
  W   V   A

The greater the power of an appliance, the more current it draws from the mains.

Typical powers	
Immersion heater	3000 W (3 kW)
Fire	3000 W (3 kW)
Kettle	2400 W (2.4 kW)
Iron	720 W
Drill	360 W
Colour TV	120 W
Table lamp	60 W
Fluorescent light	40 W

1. Do not cut right through the inner insulation when removing the outer cover.

2. Gently lever out the fuse.

3. Make sure the outer cover is gripped.

4. Wrap the wires clockwise round the studs.

Questions
1. What is the difference between AC and DC? Which type of current flows from a mains socket?
2. What is the voltage of the mains supply?
3. Why is an earth wire connected to the body of a kettle?
4. Why should switches and fuses always be placed in the live wire?
5. When does a fuse cut off current? How does it do it?
6. What equation links watts, volts and amperes? Each appliance in the chart (above right) is connected to a 240 volt supply. Use the power equation to find the current flowing through:
   (a) the kettle;   (b) the iron;   (c) the drill;   (d) the TV.

Electricity around the house

Each mains socket in a house is a part of a whole system of parallel circuits that branch off from the Electricity Board's main supply cable.

House circuits

The Electricity Board's cable into each house contains a live and a neutral wire. At the *consumer unit* or 'fuse box', these wires branch into several parallel circuits which carry current to the lights, the cooker, the immersion heater and the mains sockets. The cable for each circuit contains an earth wire, as well as a live and a neutral.

Each circuit passes through a fuse in the consumer unit. The fuse blows if the current is greater than it should be:

upstairs lighting circuit – 5 A fuse;
downstairs lighting circuit – 5 A fuse;
circuit through cooker – 30 A fuse;
circuit through immersion heater – 15 A fuse.

Mains sockets   In many houses in Europe, each mains socket is on a separate circuit with its own fuse in the consumer unit.

In many houses in Britain, all the mains sockets are connected to a *ring main* as shown in the diagram below. This is a cable which begins and ends at the consumer unit; the live, neutral and earth wires in the cable each form a long loop or 'ring' around the house. An advantage of the system is that there are two conducting paths to each socket, so thinner cable can be used.

The ring main is protected by a 30 A fuse in the consumer unit. In addition, each appliance is protected by a fuse in its plug.

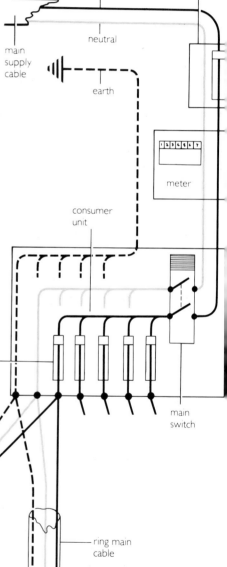

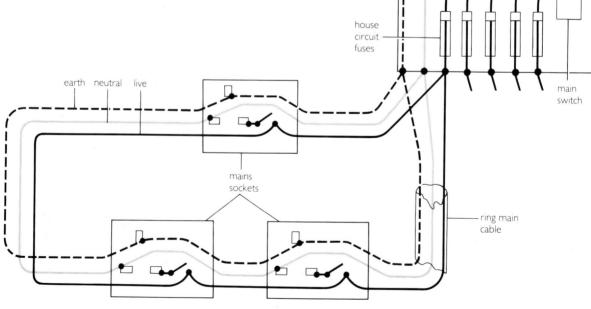

Three pin plugs

In many European countries, each appliance is sold with an unfused plug already moulded onto the end of the cable. Commonly used in Britain, is the square-pin plug on the right which you can connect to the cable yourself. This plug contains a fuse.

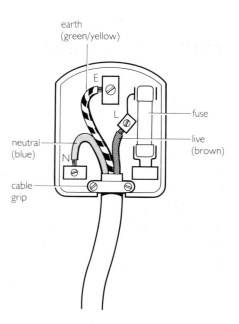

Wiring a plug   When wiring a plug, it is important to check that:
1. the three wires in the cable are connected to the correct terminals. The cable 'colour code' is shown in the diagram;
2. there are no loose strands of wire, and the cable is held firmly by the grip;
3. a fuse of the correct value is fitted.

Fuse values   Plugs are normally fitted with 3 A or 13 A fuses. The fuse value must be greater than the current which normally flows, but as close as possible to this value so that the fuse will blow before a faulty appliance can start a fire. For example:

for a kettle taking a current of 10 A – a 13 A fuse;
for a TV taking a current of 1 A – a 3 A fuse.

Changing a fuse   If the fuse in a plug blows, you must:
1. switch off at the socket and pull out the plug;
2. trace the fault;
3. fit a new fuse only when the fault has been put right.

Buying electrical energy

The Local Electricity Board charges for the electrical energy it supplies. The amount of energy supplied is measured by the *meter* between the main supply cable and the consumer unit.

The meter measures energy in kilowatt-hours (kWh).

One kilowatt-hour is the energy used by a one kilowatt appliance switched on for one hour.

One kilowatt hour is equal to 3 600 000 joules.

The greater the power of an appliance, and the longer it is switched on for, the more energy it uses. A 3 kilowatt fire switched on for 4 hours uses 12 kilowatt-hours of energy.

Questions

1. What does a 'consumer unit' contain?
2. What is a ring main? What advantage does it offer?
3. In a fused plug, which wire is connected to the fuse? What colours are used for the live, neutral and earth leads?
4. What valve fuse would you use in a plug connected to   (a) a fire taking 12 A;   (b) a food mixer taking 1 A?
5. If the fuse in a plug blows, what must you do before replacing it with a new one?
6. How much electrical energy is used by a 2 kW heater switched on for 10 hours?

5. Tighten the screws firmly.

6. Replace the fuse. Check that it fits snugly.

7. Check there are no loose strands anywhere in the plug.

8. Replace the cover.

Chemical effect of a current

Most liquids don't conduct electricity. But some do. When they do, materials start to move about and chemical changes start to happen.

Electrolysis of water

A torch bulb wired up to a 3 volt battery lights up. If one wire is cut in two and the two ends are dipped into a beaker of distilled water the bulb does not light up. This shows that pure water does not conduct electricity. However, if salt is stirred into the water, bubbles appear around the wires and the bulb lights up again. The salt solution is a good conductor. Many other salts, acids and alkalis also cause water to conduct electricity: they are called *electrolytes*.

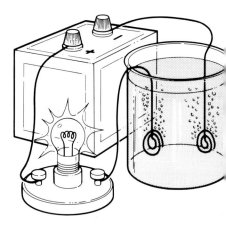

Ions as conductors   Acids, alkalis and salts split up into free-moving charged particles called *ions*, when they are dissolved in water. It is these charged particles which carry the current of electricity through the solution. The positive (+) ions are attracted towards the negative wire (the *cathode*) and move towards it. The negative (−) ions travel to the positive wire (the *anode*) to give up their extra electrons. Thus both kinds of ion are electricity carriers.

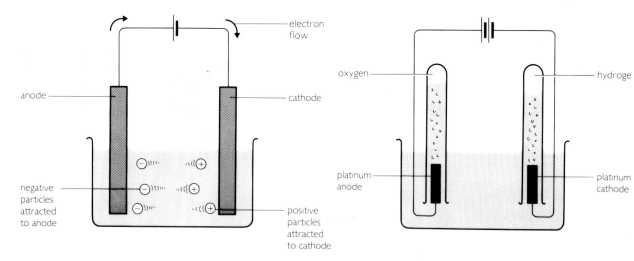

Products of electrolysis of water   The bubbles of gas given off at the anode and cathode when electricity is passed through salt solution can be collected using the apparatus shown (called a *voltameter*). The anode and cathode are small pieces of platinum metal instead of bare wires. When tested the cathode gas is found to be hydrogen and the anode gas to be oxygen. The gases are the same if small amounts of acids or alkalis are used instead of salt. The effect of the current is to split up the water into hydrogen gas and oxygen gas. The acid helps this splitting up process:

water $\xrightarrow[\text{electricity}]{\text{acid}}$ hydrogen + oxygen
$2H_2O$ $2H_2$ O_2

Uses of electrolysis

Making pure metals Many metals are extracted from their ores or *refined* by electrolysis. Pure copper, needed for electrical wiring, is produced by the electrolysis of copper sulphate solution.

A high current is used to drive positive copper ions from the solution, onto the cathode. Negative sulphate ions travel to the anode which is made of impure copper. This gradually dissolves to replace copper ions in solution and the impurities form a sludge. The electrolyte is unchanged and copper is transferred from anode to cathode.

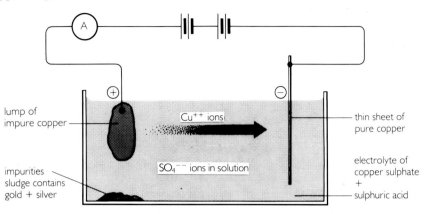

lump of impure copper

Cu^{++} ions

thin sheet of pure copper

impurities sludge contains gold + silver

SO_4^{--} ions in solution

electrolyte of copper sulphate + sulphuric acid

Electroplating If the cathode of the cell shown above were made of iron, it would still get coated with copper. It would be copper plated. Electroplating iron in this way protects it from corrosion. Layers of nickel or chromium on top of the copper give even better protection and a bright shiny finish.

To deposit a hard, even layer of metal which will not flake off, electroplating must be done very slowly. Good silver plating demands the following conditions:

The article to be plated must be clean, have a good surface and be wired as the cathode.

A very low current must be used and the temperature kept constant.

The electrolyte must have only a low concentration of silver metal (a cyanide salt is used).

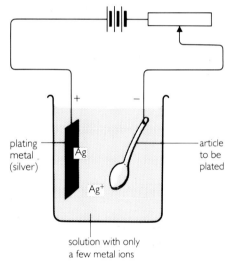

plating metal (silver)

Ag

Ag^+

article to be plated

solution with only a few metal ions

Records Electroplating plays a vital part in making records. A machine activated by sound waves cuts grooves into a plastic disc. These grooves are plated with various metals, topped with chromium. The series of metal ridges is used to stamp out copies of the original disc.

Questions

1. What is meant by the term 'electrolysis'?
2. How can you show that pure water does not conduct electricity and that salt solution does conduct electricity?
3. What is an electrolyte? What actually carries the current?
4. Name the products of electrolysing water with a little acid added.
5. Explain how impure copper is refined.
6. Why are some metals electroplated? Give three conditions necessary for successful silver plating.

Magnets

Magnetism is a mysterious force. Hang a magnet up and it always tries to point in the same direction; put it near metals and it will attract some and not others. It is known where magnetism comes from, but what it is, still isn't known.

Magnetic poles

Dip a small bar magnet into iron filings and the filings cling in clumps around the two ends of the magnet. The magnetic force pulling the filings seems to come from two points only – these points are known as the *poles* of the magnet.

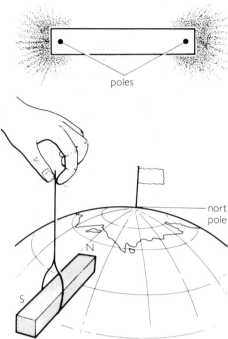

poles

N-poles and S-poles Suspend a bar magnet at its centre with a length of thread and the magnet swings round until it lies roughly north-south. The effect is used to name the poles of the magnet – you can find out more about it on page 235.

north pole

The pole at the end of the magnet pointing north is called a *north-seeking pole*, or *N-pole* for short; the pole at the end of the magnet pointing south is called a *south-seeking pole* or *S-pole* for short.

Forces between poles Bring the ends of two identical bar magnets together and there is a force between the poles:

Like poles repel each other; unlike poles attract each other.

Permanent and temporary magnets Pieces of iron and steel *become* magnets when placed next to the end of a magnet – the magnet *induces* magnetism in both metals, and each is held to the end of the magnet by the force of attraction between opposite poles.

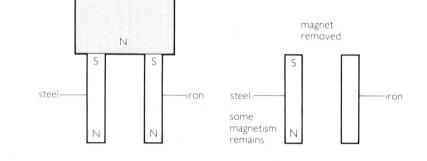

When the two pieces of metal are pulled away from the magnet, the steel keeps some of its magnetism but the iron does not. The steel has become a *permanent* magnet, the iron was only a *temporary* magnet.

Making a magnet

A piece of steel can be made into a stronger magnet by stroking it with one end of a bar magnet. The magnet must be moved along the steel over and over again *in one direction only* – the pole produced at the end of a stroke is always the opposite of the stroking pole. (A less simple but more effective way of magnetising steel is described on page 236.)

Magnetic materials

Materials which can be magnetised are called magnetic materials. They all contain one or more of the metals iron, nickel or cobalt. Steel, for example, is an alloy of iron.

Hard magnetic materials, such as steel and alcomax (a steel-like alloy), are the most difficult to magnetise but do not easily lose their magnetism. They are used to make permanent magnets.

Soft magnetic materials, such as iron and mumetal (a nickel-based alloy) are easier to magnetise but lose their magnetism easily. They are used in electromagnets (see page 236) because their magnetism can be 'switched' on and off.

Materials such as brass, copper, aluminium, lead and non-metals cannot be magnetised and are not therefore attracted to magnets.

Magnetic atoms

In a magnetic material, each atom behaves as a tiny magnet. These magnetic atoms line up with each other in groups called *domains* which point in all directions in an unmagnetised material.

When a material is magnetised, the magnetic atoms are pulled round until all the domains point in the same general direction. Throughout most of the material, the poles of each magnetic atom cancel out the effects of opposite poles nearby. At each end of the material however, the poles of the atoms act together to produce the effect of a single N- or S-pole.

Heat a magnet to red heat or hammer it repeatedly and the domains are thrown out of line. The material ceases to be a magnet.

Questions

1. What is meant by the 'N-pole' of a magnet?
2. Name two poles that (a) attract each other; (b) repel each other.
3. What is happening in the diagram on the right? Copy the diagram and complete it.
4. Which is more suitable for making a permanent magnet, steel or iron? Why? What other material could be used?
5. Which of the following materials are not attracted to a magnet: iron, steel, aluminium, copper, wood, nickel?
6. How could you make a magnet lose its magnetism?

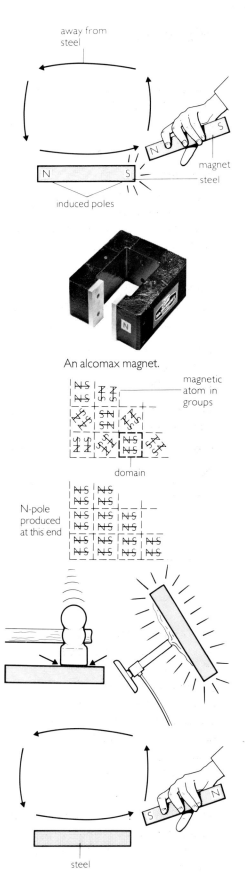

An alcomax magnet.

Magnetic fields

A bar magnet produces its strongest effects near its poles but its magnetic influence is spread throughout the whole of the space around it. And there is one magnet whose influence reaches just about every place on Earth . . .

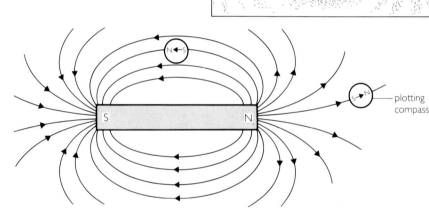

The field around a bar magnet
Sprinkle iron filings on to a piece of paper covering a bar magnet, and they will form a pattern like the one shown on the right. The filings are turned into position by forces from the poles of the magnet. The region in which these forces act is called a *magnetic field* and it can be studied with the aid of a small plotting compass.

The plotting compass   This contains a tiny pointed magnet (called a 'needle') supported on a spindle through its middle so that it can turn freely. Placed close to a magnet as in the diagram, the needle is turned by forces between its poles and the poles of the magnet. The forces pull the unlike poles towards each other and push the like poles away from each other, the needle coming to rest when the turning effects of all the forces on it are in balance.

The magnetic field around the magnet can be drawn as a series of lines running from its N-pole to its S-pole. Each section of line shows how a small compass needle would lie if placed in that region; the arrows show the direction in which the N-pole end of the needle would point.

plotting compass

Fields between magnets
The diagrams show the magnetic fields produced when the ends of two magnets are put close together. A compass needle placed at the point X is not turned by the magnets. At this point, the poles of the magnets have equal but opposite effects, so the field from one magnet cancels out the field from the other. X is called a *neutral point*.

A plotting compass.

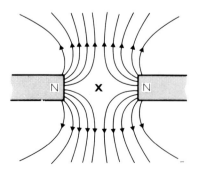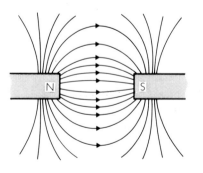

The Earth's magnetic field

The Earth itself produces a magnetic field. Whatever the actual cause of its magnetism, the Earth behaves as if a huge bar magnet were buried through its centre.

If there are no other magnets around to affect it, a compass needle turns into line with the Earth's magnetic field. Its N-pole is pulled towards the Earth's magnetic S-pole – any bar magnet which is free to turn behaves in the same way.

The N-pole end of the compass needle actually points in the general direction of the North Pole. The Earth's magnetic S-pole lies under the geographical North Pole.

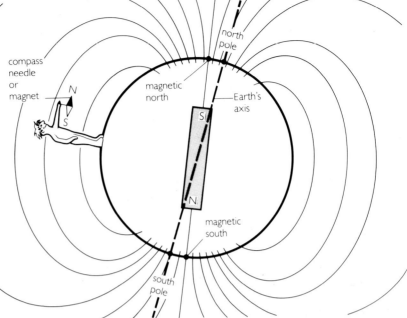

Angle of declination   At most places on Earth, a compass needle doesn't give the direction of true north because the Earth's magnetic poles are not quite in line with its north-south axis.

The angle between true north and the compass needle direction (magnetic north) is called the *angle of declination*, and it varies in value depending on where on the Earth's surface you happen to be. Navigators must make allowance for the local angle of declination when using a compass reading to set a course.

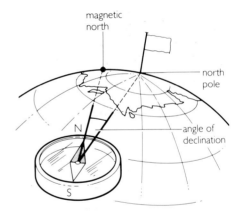

Questions

1. How could you show that there is a magnetic field around a bar magnet?
2. Magnetic fields are often drawn as a series of lines. What does each section of line show? What do the arrows show?
3. Copy the diagram on the right and draw lines to show the magnetic field. Draw on a neutral point if there is one. What is a neutral point?
4. What is a compass needle? How does a compass needle show that the Earth's S-pole lies towards the North Pole?
5. Why does a compass needle not point true north?
6. What is meant by 'angle of declination'?

235

Electromagnets

You can make magnets in which the magnetism comes and goes at the flick of a switch. It's all made possible because every electric current produces a magnetic field.

Electromagnets
Pass a current through a coil of wire and the coil acts as a magnet. Wind the coil on an iron *core* and the magnetic field becomes much stronger – the diagram shows a simple electromagnet made in this way. To produce an even stronger magnetic field, you could:

pass a larger current through the coil;
use more turns of wire in the coil.

Core materials   If the core of an electromagnet is made of soft iron, its magnetic field will vanish when the current through the coils is switched off. If the core is made of steel, it remains magnetised – the principle is used in making permanent magnets.

Magnetic pole   The diagram on the right shows you how to work out which pole of an electromagnet is which. Gripping the coil as shown, your thumb points towards the N-pole.

Several uses of electromagnets are described here.

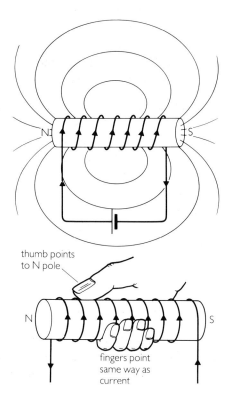

thumb points to N pole

fingers point same way as current

The electric bell
An electric bell contains an electromagnet that switches itself off and on very rapidly, moving the bell hammer as it does so:

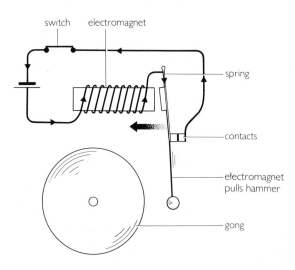

switch   electromagnet
spring
contacts
electromagnet pulls hammer
gong

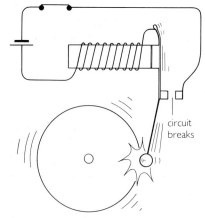

circuit breaks

When the bell switch is pressed, current flows through the electromagnet and the hammer is pulled across to strike the gong. The movement pulls the contacts apart, which cuts power to the electromagnet – the hammer springs back, the contacts close again and the process repeats itself until the bell switch is released.

Relays

Problem A car starter motor takes a current of over 100 A and needs very thick cable to connect it to a car battery. But it has to be switched on by a lightweight ignition switch, connected to thin cable. This switch can't handle the high current.

Solution Use an electromagnetic switch, or *relay*. With a relay, a large current in one circuit can be switched ON or OFF by a small current in another circuit.

The diagram shows how a relay can be used to switch on an electric motor. When the switch in the input circuit is closed, the electromagnet comes ON and pulls the iron armature towards it. This closes the contacts, so the motor in the output circuit is switched ON.

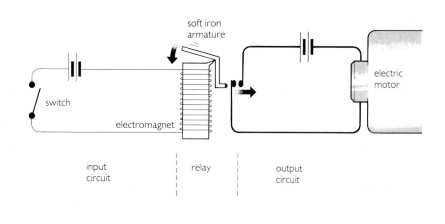

There are two types of relay:

Normally open relay With this, the output circuit is switched ON when the input circuit is switched ON. For example: a car starter motor is switched ON when the ignition switch is turned ON by a key.

Normally closed relay With this, the output circuit is switched OFF when the input circuit is switched ON. For example: a workshop lathe is switched OFF when a red 'stop' button is pushed ON.

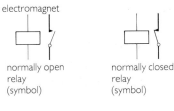

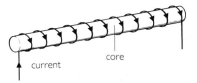

Questions

1. What material would you use for the core of an electromagnet?
2. How could you make the field from an electromagnet stronger?
3. Copy diagram A. Mark in the N- and S-magnetic poles.
4. What makes the hammer of an electric bell hit the gong? Why does the hammer move back afterwards?
5. A shop alarm system is shown in diagram B. It is switched on by closing the switches X and Y. If anyone tries to steal a radio, the tamper circuit is broken and the alarm bell rings.
   (a) What type of relay is being used?
   (b) Why will the alarm ring if switch X only is closed?
   (c) Why will the alarm stop ringing if switch Y is closed?
   (d) Why will the alarm bell start ringing again if the tamper circuit is broken?

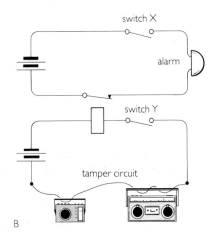

Using magnetism to make electricity

An electric current produces a magnetic field – but a magnetic field can also be used to produce an electric current.

Induced voltage and current

The coil on the right is connected to a sensitive meter. The meter shows that a small current flows through the coil whenever the magnet is moved – the moving magnetic field produces a voltage in the coil, and this *induced* voltage makes a current flow because the coil is in a complete circuit.

Current direction   Move the magnet into the coil and the current flows one way; move it out of the coil and the current flows in the opposite direction. In both cases, the induced current makes the coil into an electromagnet which tries to stop the movement of the magnet – repelling the magnet as you push it in and attracting it as you pull it out. In producing electrical energy you have to do work to move the magnet.

Strength of induced voltage   A voltage is only induced in the coil when the magnet is moving. A higher voltage is induced if:

the magnet is moved more rapidly;
a stronger magnet is used;
more turns are wound on the coil.

The induction coil

Switching off an electromagnet inside a coil has the same effect as moving a magnet very rapidly out of the coil. Both induce a high voltage. This is the principle behind the *induction coil* used in many cars to produce high voltage sparks in the plugs.

An induction or *ignition* coil is actually two coils wound one inside the other, around a soft iron core, as shown in the diagram:

The primary coil and core are an electromagnet which switches off every time one of the bumps (cams) on the spinning distributor shaft pushes apart the contacts in the primary circuit. In electronic ignition, a transistor is used instead of moving contacts.

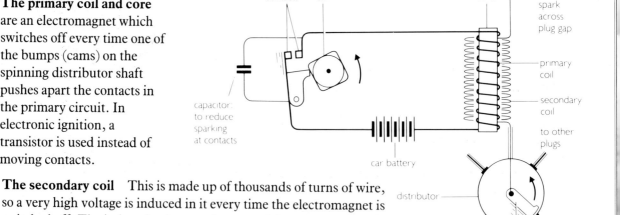

The secondary coil   This is made up of thousands of turns of wire, so a very high voltage is induced in it every time the electromagnet is switched off. The induced voltage makes a spark jump across the air gap in one of the plugs – each plug is linked to the coil in turn by the spinning rotor arm in the distributor.

Transformers

A transformer is a device which changes an AC voltage to a higher or lower AC voltage. Like an induction coil, it has a primary coil, a secondary coil, and a core.

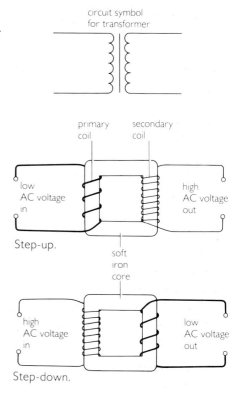

circuit symbol for transformer

primary coil   secondary coil

low AC voltage in

high AC voltage out

Step-up.

soft iron core

The primary coil and core   When the primary coil is connected to an AC supply, current surges backwards and forwards through it 50 times every second (see page 226). As the current rises, falls and reverses over and over again, the coil and core act as an electromagnet which is being switched on and off very rapidly.

Secondary coil   The rapid 'switching' of the primary coil induces an AC voltage in the secondary coil. The value of this voltage can be worked out using the equation:

$$\frac{\textbf{voltage across secondary coil}}{\textbf{voltage across primary coil}} = \frac{\textbf{turns in secondary coil}}{\textbf{turns in primary coil}}$$

Step-up transformer   This has more turns in the secondary coil than in the primary. It gives a higher voltage than is put into it.

Step-down transformer   This has less turns in the secondary coil than it has in the primary. It gives a lower voltage than is put in.

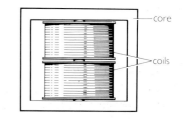

high AC voltage in

low AC voltage out

Step-down.

Power in a transformer   A completely efficient transformer gives out as much power (energy every second) as is put into it:

$$\frac{\textbf{volts} \times \textbf{amperes}}{\textbf{in primary coil}} = \frac{\textbf{volts} \times \textbf{amperes}}{\textbf{in secondary coil}}$$

Transformer construction   To be completely efficient, a transformer must waste none of the magnetic field produced in the primary coil and none of the power put into it. The arrangement of coils and core shown on the right is designed to 'trap' as much of the magnetic field as possible; the soft iron core is *laminated* (made of thin insulated sheets) so that power is not wasted by currents induced in the core itself.

core

coils

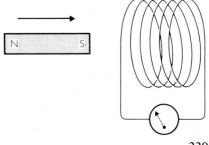

Questions

1. The magnet on the right is pushed into the coil. When does the current stop flowing? What happens when the magnet is pulled out of the coil?
2. In a car ignition coil, why does the magnetic field suddenly disappear? What happens when it does? How is such a high voltage produced?
3. A transformer has 1000 turns in its primary coil and 100 turns in its secondary. The primary coil is connected to a 200 volt AC supply. What voltage is produced across the secondary coil?
4. What is the difference between a step-up and a step-down transformer? Why are the cores of transformers laminated?

N   S

Motors and generators

Pass a current through an electric motor and the motor starts to turn. A generator uses the reverse effect: turn a generator and it can produce an electric current.

A simple DC electric motor

The electric motor in the diagram contains a coil in a magnetic field. The coil spins round when direct current (DC) flows through it – in this case, the current is supplied by a battery.

Two carbon contacts, known as *brushes*, connect the battery leads to the turning coil. The brushes are pushed against the two halves of the copper split ring or *commutator* by small springs – the commutator is fixed to the coil and turns round with it.

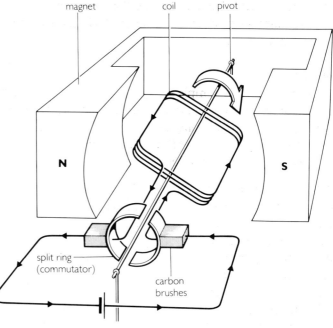

How the motor works When a current goes through it, the coil acts as a short, wide electromagnet with an N-pole and an S-pole. These poles are pulled towards the opposite poles of the permanent magnet. The coil would tend to come to rest once its N- and S-poles were as close as possible to the S- and N-poles of the permanent magnet. But just as it is reaching this point, something happens . . .

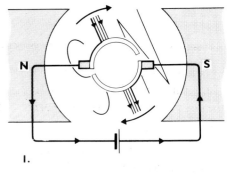

The two splits in the split ring are located so that just at that instant, they pass under the brushes, and the brushes 'change halves' – they now press against the opposite half rings. This means that the current flows round the coil in the opposite direction – the coil's N- and S-poles become S- and N-poles respectively. Instead of being very close to the 'coming to rest' point, they are now almost exactly opposite it! So round the coil goes, and as it reaches the 'come to rest' point, something happens . . . In this way, the coil continues to revolve, round and round.

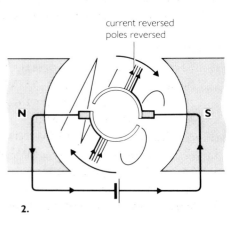

Practical motors Practical motors usually have several coils, each connected to its own pair of commutator pieces. This makes the motor more powerful and smooths out the uneven turning effects on each coil. To make the motor even more powerful, many turns of wire are used in the coils and they are wound on a soft iron *armature* – this becomes magnetised and increases the strength of the magnetic field through the coils.

A simple DC generator

Spin round the coil of a simple DC motor and the motor becomes a *generator* – the spinning coil has a voltage induced in it and this can push a current round a circuit.

Any current induced in the spinning coil flows backwards and forwards as the coil faces first one way and then the other. The generator gives out 'one way' direct current (DC) because of the action of the commutator.

Alternators

Unlike a DC generator, the simple AC generator or *alternator* in the diagram does not have a commutator. Generated current flows from the coil through two complete rings *(slip rings)* which rub against the brushes. As the coil spins, current flows backwards and forwards through the bulb in the outside circuit.

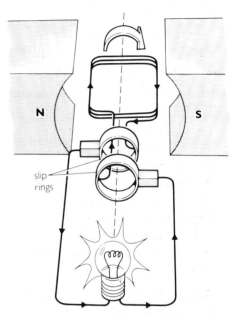

The alternating current (AC) through the bulb is shown in the graph on the right. When the coil is horizontal as in the diagram, it is moving most rapidly through the magnetic field – this is when the voltage generated is highest and most current flows through the bulb.

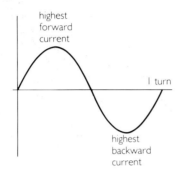

Voltage output   The alternator generates a higher voltage if:

its coil is rotated more rapidly;
a larger number of turns is used in the coil;
the coil is wound on a soft iron armature;
a stronger magnet is used.

Alternators in cars   Most cars are fitted with alternators. In a car alternator, AC is generated in fixed coils placed around an electromagnet which is turned by the engine. The AC is changed to DC by devices called diodes and used to charge the battery.

Questions

1. In a simple DC motor, what is the purpose of   (a) the brushes;   (b) the commutator? Why does the coil turn when a current flows through it? Give three ways in which the motor could be made more powerful.
2. What happens if you spin the coils of a simple DC motor?
3. What does an alternator generate? What are slip rings?
4. When is the voltage across an alternator coil greatest?
5. How can the voltage from a simple alternator be increased?
6. Give an example of the practical use of alternators.

Generating and transmitting electricity

Mains electricity is generated by huge alternators sited in power stations and sent across the country mainly through overhead cables. There are many transformers between a power station and the mains points in your house, and your power could come from any power station in the country.

Power stations

The diagram shows in simplified form the layout of a typical power station. The alternator is one of several on the site, all driven by huge turbines which are spun round by the force of high pressure steam. The steam comes from water heated in a boiler by burning coal, gas or oil or by using the heat from a nuclear reactor.

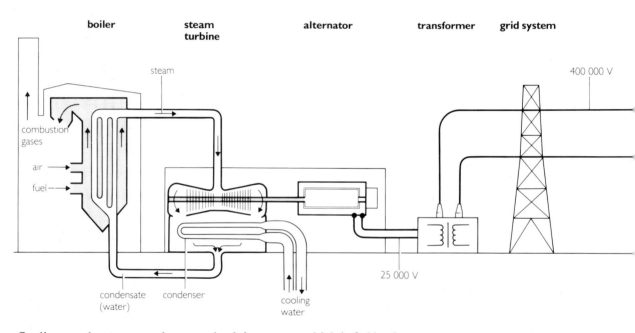

Cooling condensers turn the steam back into water which is fed back to the boiler for reheating. Vast quantities of cooling water are needed for the condensers, so the power station will if possible be sited near a river or the sea.

Power to the cables

Electric power from the alternator is fed to a huge transformer where the voltage is stepped up from 25 000 volts to 400 000 volts. At this higher voltage, power is carried across country by overhead cables.

The voltage increase is made in order to reduce the current which must be carried by the overhead cables:

As power in watts is given by *voltage × current*, any increase in voltage must produce a drop in current if the transformer is to give out the same power (energy every second) as is put into it. Power leaving the transformer is carried by a smaller flow of electrons, though the energy of each individual electron is greater.

By reducing the current through the cables in this way, the transformer enables thinner and lighter cables to be used. One of the advantages of generating and transmitting AC is that voltages can be stepped up with a transformer to reduce current – changing DC voltages is very difficult.

The Grid

The overhead cables from the power station feed power to a nationwide supply network called the *Grid*. Since over 180 power stations feed power to the Grid, the system is very flexible. Power stations in areas where the demand for electricity is low can be used to supply an area where the demand is high.

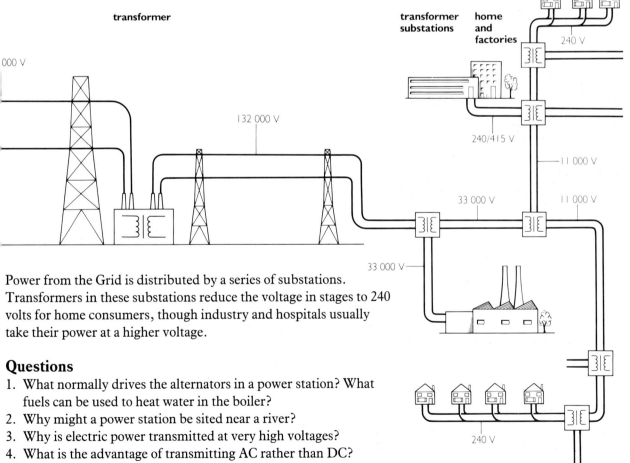

Power from the Grid is distributed by a series of substations. Transformers in these substations reduce the voltage in stages to 240 volts for home consumers, though industry and hospitals usually take their power at a higher voltage.

Questions

1. What normally drives the alternators in a power station? What fuels can be used to heat water in the boiler?
2. Why might a power station be sited near a river?
3. Why is electric power transmitted at very high voltages?
4. What is the advantage of transmitting AC rather than DC?
5. What is the main advantage of the Grid network?
6. What are substations for?

243

Nuclear energy

Nuclear power stations use uranium as fuel. But the uranium doesn't burn. Its energy is released in a completely different way.

Radioactivity

The photograph shows the fuel rods used in a nuclear power station. The uranium in the rods is *radioactive*. This means that its atoms are unstable. In time, the nucleus of each atom breaks up and shoots out a tiny high-speed particle.

The break-up of unstable atoms is called *radioactive decay*. The particles shot out are called *nuclear radiation*.

Fuel rods from nuclear power station.

Radioactive atoms aren't only found in nuclear power stations. There are tiny amounts in the ground, the air, and even in living things. This is because most elements are a mixture of isotopes, and some isotopes are unstable. The chart gives some examples.

Isotopes		Found in
Stable	*Unstable, radioactive*	
carbon-12 carbon-13	carbon-14	air, plants animals
potassium-39 potassium-41	potassium-40	rocks, plants sea water
	uranium-235 uranium-238	rocks

Radioactivity produces heat. When unstable atoms break up, their nuclear radiation hits other atoms and makes them move faster. So nuclear energy is changed into heat energy. The high temperatures deep in the Earth are due to radioactivity in the rocks.

Heat from uranium

Natural uranium is mainly uranium-238, with small amounts of uranium-235 mixed in. The two isotopes are very difficult to separate.

Both types of uranium break up naturally – but slowly. However, uranium-235 atoms can be made to break up quickly. They can be split by neutrons in a nuclear reactor. The splitting is called *fission*. It works like this:

A uranium-235 atom is hit by a neutron. The atom breaks up, releases energy and shoots out more neutrons. These hit other uranium-235 atoms and make them split – and so on. The result is a *chain reaction*. Huge amounts of heat are released very quickly.

Chain reactions won't work without uranium-238. Atoms of uranium-238 absorb neutrons without splitting.

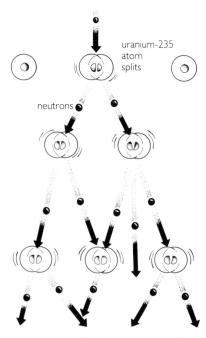
uranium-235 atom splits

neutrons

Fission of uranium-235.

A nuclear power station

A nuclear power station uses heat to make steam – just like coal, oil or gas-burning power stations. But the heat is produced in a nuclear reactor.

The diagram shows an *Advanced Gas Reactor*. It contains:

Nuclear fuel rods   These are made from uranium dioxide. They contain natural uranium with extra uranium-235 mixed in. 1 kg of this 'enriched' uranium fuel gives as much energy as 55 tonnes of coal.

A core made of graphite blocks   The graphite is a *moderator* – it slows down neutrons released by fission. If the neutrons aren't slowed they don't cause fission. Instead they escape, or are absorbed by uranium-238 atoms.

Control rods   These are raised or lowered to control the rate of fission. They are made of boron, which absorbs neutrons. If the rods are raised, more neutrons can cause fission. So the reactor temperature rises. If the rods are fully lowered, the chain reaction stops and the reactor cools down.

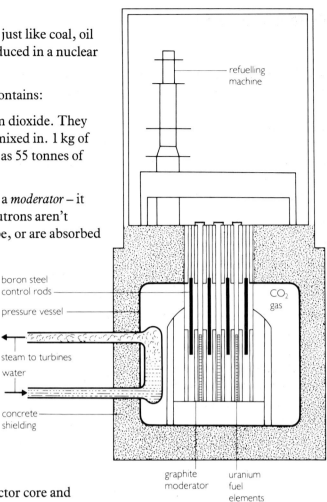

Carbon dioxide gas.   This flows through the reactor core and carries heat to the boiler.

The whole reactor is inside a steel pressure vessel. This is surrounded by thick concrete to absorb the radiation.

Questions

1. Using words from the following list, copy and complete the sentences below:
   unstable   nuclear   neutrons   fission   isotopes
   (a) Radioactive atoms are . . . . When they break up, they give off . . . . radiation.
   (b) Uranium-235 atoms will split when hit by . . . . This is called . . . .
2. *boron   graphite   uranium   concrete   steel*
   In a reactor, which of these materials is used:
   (a) as a moderator?   (b) to absorb radiation?   (c) in the fuel rods?   (d) in the control rods?
3. Explain why, in a nuclear reactor:
   (a) a moderator is needed
   (b) carbon dioxide is pumped through the reactor core
   (c) the chain reaction stops if the control rods are fully lowered.

A nuclear power station in Wales.

Nuclear radiation

When the fuel rods in a nuclear reactor are exhausted, they have to be removed. But the waste material in them is still radioactive – and very dangerous.

Danger! Radiation

The steel flasks in the photograph are carrying radioactive waste from a nuclear reactor. If any of it leaked out, it could contaminate the air, crops, and the local water supply. And many deaths could result.

Nuclear radiation can damage or destroy vital body cells. An overdose can cause cancer or fatal radiation sickness. Radioactive gas and dust is especially dangerous because it can be taken into the body with air, food or drink. Once absorbed, it can't be removed, and its radiation causes cell damage deep in the body.

Types of nuclear radiation

Radioactive materials give out three main types of nuclear radiation: *alpha particles*, *beta particles* and *gamma rays*:

Waste from a nuclear reactor.

Type	Alpha particles	Beta particles	Gamma rays
	each particle is 2 protons + 2 neutrons	each particle is an electron	electromagnetic waves, similar to X-rays
charge	+	−	no charge
penetrating effect	stopped by thick sheet of paper, or skin	stopped by 5 mm of aluminium	reduced by lead, but not stopped

Nuclear radiation can be detected using a *Geiger-Müller tube* like the one in the photograph. If, say, a stream of alpha particles enters the tube, the tube gives off a series of electrical pulses – one for each particle. These are counted by the meter. If five particles enter the tube every second, the meter shows a reading of five counts per second. This is called the *count-rate*. The more the radiation, the higher the count-rate.

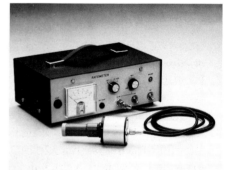

A Geiger-Müller tube measures nuclear radiation.

Half-life

The radiation from a radioactive material gets less as time goes on, but it never completely disappears. This is how the count-rate readings changed when a Geiger-Müller tube was held near some radioactive radon-222 gas:

start	32 counts per second
after 4 days	16 counts per second
after 8 days	8 counts per second
after 16 days	4 counts per second and so on ...

The amount of radiation halved every four days.
Radon-222 has a *half-life* of four days.
The chart shows the half-lives of some other radioactive materials.

Material	Half-life in years
strontium-90	28
radium-226	1620
carbon-14	5770
plutonium-239	24 360
uranium-235	700 000 000

Decay hazards

When radioactive atoms break up, they change into atoms of completely different materials. These materials are called *decay products*. The decay products from Britain's nuclear reactors are sent to the reprocessing plant at Sellafield to be separated. Many of them are highly radioactive. And many have very long half-lives. It will be hundreds of years before their radiation has dropped to a safe level.

Sellafield nuclear reprocessing plant, Cumbria.

Some decay products are especially dangerous. *Strontium-90* and *iodine-131* are easily absorbed by the body. Strontium becomes concentrated in the bones, iodine in the thyroid gland. *Plutonium-239* is the most dangerous substance of all. Breathed in as dust, the smallest amount can kill. Plutonium is used in nuclear weapons, and as a fuel in certain types of nuclear power station.

Questions

1. Which type of nuclear radiation:
   (a) has no charge? (b) can penetrate lead? (c) is made up of electrons? (d) has a − charge? (e) is stopped by a thick sheet of paper?
2. Explain why:
   (a) nuclear radiation is dangerous.
   (b) radioactive gas and dust is especially dangerous.
3. Look at the table of half-lives on this page. If samples of strontium-90 and radium-226 both give the same count-rate today, which would give the highest reading in 10 years' time?
4. The table on the right shows how the count-rate from some iodine-131 changed. What is the half-life of iodine-131?

Time in days	Count-rate in counts per second
0	240
4	170
8	120
12	85

Further questions

1 In an atom, what particles carry
   a negative (−) charge?
   b positive (+) charge?
   c no charge?

2 *Aluminium, copper, iron, carbon, water, glass, polythene, PVC*
From the above list, select:
a the good conductors of electricity
b the poor conductors of electricity
c the insulators

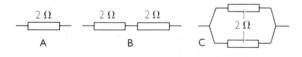

3 Copy the diagram above. Mark in the + and the − terminals of the cell, and the direction in which the electrons flow.

4 Give the approximate voltage of
   a a small torch battery
   b a car battery
   c an AC mains socket

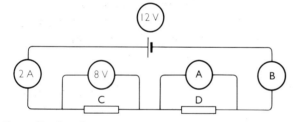

5 **a** In the diagram above, what type of meter is **A** and what type of meter is **B**?
   b What will be the readings on **A** and **B**?
   c Using the values given for the voltage across **C** and the current through it, find the resistance of **C**.
   d What is the resistance of **D**?

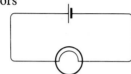

6 **a** Which of the above arrangements offers most resistance to an electric current?
   b Which offers least resistance?

7 **a** Draw a diagram to show how you would connect three 2 V bulbs in series with a battery. What should the voltage of the battery be?
   b Redraw the diagram to show the three bulbs connected in parallel. What battery voltage is required in this case?

8 An electric fire element is connected to the 240 V mains. If the power of the element is 1200 W what current flows?
Would you connect a 3 A or a 13 A fuse to the plug fitted to the fire? Give reasons for your answers.

9 The diagram below shows a fused square-pin mains plug:

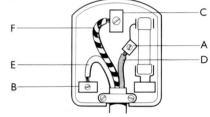

Name the terminals marked A, B and C. Give the colours used for the insulation on the wires D, E and F.

10 In the experiment below, an electric current is being passed through water with a small amount of sulphuric acid added.

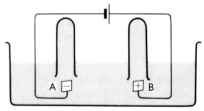

a Which electrode is the cathode?
b What is the other electrode called?
c What gas collects above electrode **A**?
d What gas collects above electrode **B**?
e Above which electrode will most gas collect? Why?

11 *Iron, copper, brass, nickel, aluminium, steel, carbon, cobalt, glass*

In the above list:

   a Which materials are magnetic?

   b Which materials are non-magnetic?

12 Copy and complete the above diagram to show the field around the magnet.

13 **a** Why is the core of an electromagnet made of iron rather than steel?

   b What will happen to the magnet in the diagram above when the electromagnet is switched on?

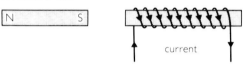

   c Give two examples of the practical use of electromagnets.

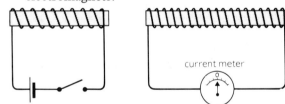

14 **a** In the above experiment, what happens when the electromagnet is switched on?

   b What happens when the electromagnet is switched off?

   c What practical use can be made of the principle demonstrated in the experiment above?

15 **a** In circuit diagrams, what symbol is used to represent a transformer?

   b Write down the equation linking the voltages across the primary and secondary coils of a transformer and the numbers of turns on the coils.

   c The primary coil of a transformer is to be connected to the 240 V mains, the secondary coil to a 12 V light bulb. If there are 2000 turns in the primary coil, how many turns must there be in the secondary coil?

16 The diagram below shows the top view of a simple electric motor:

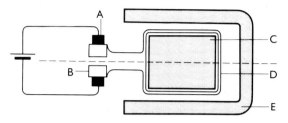

Name the parts labelled **A**, **B**, **C**, **D** and **E**.   ALSEB

17 The diagram below shows in simplified form the layout of a power station:

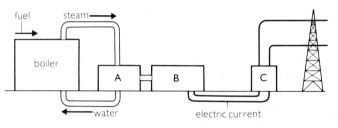

   a What happens in **A**?

   b What is **B**?

   c What is **C**? What is its purpose?

   d What must happen to the power passing through the overhead cables before it can be supplied to your home?

18 Imagine you are trying to find out if a fuse has 'blown'. You are given the following pieces of apparatus.

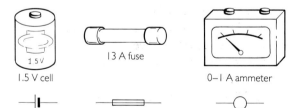

   a Draw a circuit diagram to show how you would connect these components.

   b Explain how you would tell whether or not the fuse had blown.

   c If the fuse was rated 100 mA you would need an extra component in your circuit. **i** Why is this? **ii** What extra component would you need?   SEG/SWEB

19 From the following list of components, write down the one which best fits each of the sentences given below. Each may be used once, more than once or not at all.

diode npn transistor LED LDR thermistor relay

a The . . . . normally has three terminals called the emitter, base and collector.

b The resistance of a . . . . decreases a lot as the temperature increases.

c The resistance of a . . . . decreases a lot as the light intensity increases.

d A . . . . is an electromagnetic switch which allows a larger current to be switched on by a smaller current.

NEA

20 A new type of light bulb has recently been invented. It produces the same amount of light as an ordinary 100 W bulb, but only uses 25 W of electrical power. It is expected to last for 5000 hours.

a How many kilowatt-hours does a 100 W lamp use in 5000 hours?

b How much will this cost? The electricity board charges 5 pence for 1 kilowatt-hour.

c Both bulbs produce the same amount of light energy but the old type uses more electrical energy. Explain what happens to this extra energy.

SEG/SWEB

21 a The diagram below shows a temperature controlled transistor switch.

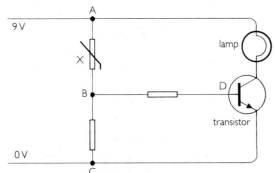

i Name the device labelled X.   **ii** State what happens to the resistance of X when the temperature rises.   **iii** Explain carefully how the change of resistance affects the circuit.

b Draw a labelled diagram of a light-controlled transistor switch which will operate when the conditions change from light to dark.

LEAG

22 The diagram below shows the stages of generating electricity in a nuclear power station.

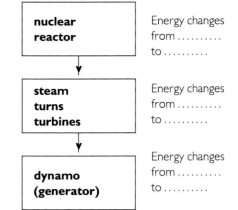

a Copy the diagram and write down the main energy changes for each stage.

b Only about ¼ of the steam's energy changes into electricity.

i In what form does the rest of the energy leave the power station?   **ii** Where does this energy normally end up?

c The nuclear reactor has concrete walls several metres thick all around it. Why is this?

d Nuclear fuels produce energy when fission occurs. Use the diagram below to explain what happens during nuclear fission.

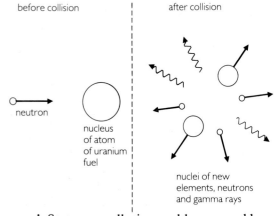

e   **i** State one pollution problem caused by nuclear power stations.   **ii** State one way of overcoming this problem.

SEG/SWEB

Answers to numerical questions

page 71   **4.** 0 °C **5.** 100 °C **6.** −183 °C
              7. 273 °C

page 81   **5.** Max 30 °C, Min 0 °C

page 89   **1.** 45 °C **2.** 2 m behind

page 93   **2.** 3 cm

page 103   **6.** 320 m/s

page 105   **5.** 320 m/s

page 107   **2.** 400 Hz, 800 Hz

pages   **3. a)** 0 °C **b)** 100 °C **c)** −273 °C
108–110   **d)** 23 °C **e)** 37 °C
          19. a) A: 1 cm, B: ½ cm **b)** 4 Hz

page 113   **2. a)** 30 m/s **b)** 30 m **c)** 180 m
            d) 8 s
          4. a) 40 m/s **b)** 20 s **c)** 50 s
            d) 40 m/s, 2 m/s, 2 m/s^2

page 115   **6.** 1000

page 117   **3.** 30 N **4.** 4 m/s^2

page 119   **1.** 30 N

page 121   **3.** 6 N

page 123   **1.** 60 J

page 125   **4.** 4800 J

page 133   **4.** 2m, 2 **6.** 2, 100 N

page 139   **3.** 30 cm^3, 10 cm^3

page 141   **1.** 12 Pa, 2 Pa

page 143   **2.** 6000 N, 6000 N

page 145   **4. a)** 100 000 Pa **b)** 760 mm Hg

pages   **7.** 40 J **14.** 2400 kg
148–150   **15. a)** 50 m^3 **b)** 75 kg
          17. b) 77 cm
          21. a) 20 s **b)** 30 m/s **c)** 30 s
            d) 3 m/s^2
          23. b) 2 m^3 **25. d)** 3100 kJ
          27. a) 400 N **b)** 600 Nm **c)** 600 N
            d) 60 kg

page 213   **2.** 11

page 217   **2. c)** 2 A **d)** 4 V
          3. 1 A, 0.5 A, 2.5 A

page 219   **3.** 6 V, 1 A

page 227   **6. a)** 10 A **b)** 3 A **c)** 1.5 A
            d) 0.5 A

page 229   **4. a)** 13 A **b)** 3 A **6.** 20 kWh

page 239   **3.** 20 V

page 247   **4.** 8 days

pages   **5. b)** A: 4 V, B: 2 A **c)** 4 Ω **d)** 2 Ω
248–250   **7. a)** 6 V **b)** 2 V
          8. 5 A, 13 A fuse
          15. c) 100 turns **d)** 36 W
          20. a) 500 kWh **b)** £25

Acknowledgements

Allsport: pp.112, 147; Heather Angel: p.201; Anglian Water Authority: p.27; Associated Press: p.119; Barnaby's Picture Library: pp.11, 21, 37, 40, 42, 45, 65, 119, 183; Berger: p.59; Besam/ Spurgeon Walker Associates: p.224; Brian Beckett: p.159; Boosey & Hawkes: p.107; Paul Brierley: p.55, rear cover; British Airways: p.146; British Gas Corporation: p.30; British Gypsum: p.41; British Leyland: pp.47, 131; British Nuclear Fuels plc: pp.246, 247; BOC: pp.28, 31; BP: p.61; British Rail: pp.76, 78; British Steel: pp.54, 57; British Tourist Authority: p.26; Cadbury-Typhoo: p.21; Camera Press: pp.12, 13, 23, 49, 56, 69, 81, 111, 151, 176, 185, 211; Cull Photographic: pp.58, 205, 206; Electrolux: p.128; Farmers Weekly: p.175; Fox: pp.43, 43, 48, 83, 103; Frank Lane Picture Agency: pp.146, 181; Griffin & George: p.146; Phillip Harris: pp.24, 156, 191, 193, 197, 199, 217; Hills & Harris: pp.39, 46, 50, 51; Honda UK Ltd: p.149; ICI: pp.52, 63; Institute of Geological Sciences: pp.9, 10; Keystone Press Agency: pp.8, 33; Lead Development Association: p.56; John Lundie: p.212; MK Electric: p.229; NASA: p.5; North of Scotland Hydro-Electric Board: p.136; National Coal Board: p.60; Olympus: p.94; Panos Pictures/Marcus Santilli: p.178; Phillips Electronics: p.222; Picturepoint: pp.178, 179, 179; Popperfoto: p.23; Radio Spares: p.102; Ann Ronan: p.25; Royal Astronomical Society: p.203; Sarsons Ltd: p.38; The Science Museum: p.10; Chris Shaw: p.71; Shell: p.137; Space Frontiers: p.118; Sparklets International: p.33; St. Ivel Ltd: p.204; Sussex Police: p.112; Jeffrey Tabberner: pp.23, 38, 73, 95, 115, 116, 227; Topham Picture Library: pp.147, 181; Unilever Research Laboratory: p.64; United Kingdom Atomic Energy Authority: p.244; US Naval Observatory: p.6; James Webb: pp.154, 155, 156, 157, 161, 172, 184, 187, 195, 203

Additional photographs by Chris Honeywell
Additional diagrams by Illustra Graphics Limited

Useful information

Useful chemicals

Main chemical present	Formula	Common name
aluminium oxide	$Al_2O_3 . 2H_2O$	bauxite
ammonium chloride	NH_4Cl	
ammonium nitrate	NH_4NO_3	nitram
butane	C_4H_{10}	camping gas
calcium carbonate	$CaCO_3$	limestone, marble, chalk
calcium hydroxide	$Ca(OH)_2$	limewater, slaked lime
calcium oxide	CaO	quicklime
calcium sulphate	$CaSO_4 . 2H_2O$	plaster (gypsum)
carbon	C	soot, charcoal, coke, graphite, diamond
copper and nickel	Cu, Ni	coins
copper and tin	Cu, Sn	bronze
copper and zinc	Cu, Zn	brass
copper(II) sulphate	$CuSO_4 . 5H_2O$	
ethane-1,2-diol	$C_2H_4(OH)_2$	antifreeze
ethanoic (acetic) acid	CH_3COOH	vinegar
ethanol	C_2H_5OH	alcohol
ethanol and methanol	C_2H_5OH, CH_4OH	methylated spirits
hydrochloric acid	HCl	
hydrogen oxide	H_2O	water
iron	Fe	steel
iron and chromium	Fe, Cr	stainless steel
iron(III) oxide	Fe_2O_3	haematite (iron ore)
iron(II) sulphate	$FeSO_4 . 7H_2O$	
iron sulphide	FeS_2	iron pyrites
lead and tin	Pb, Sn	solder
lead oxide	Pb_3O_4	red lead
magnesium hydroxide	$Mg(OH)_2$	magnesia
magnesium sulphate	$MgSO_4 . 7H_2O$	epsom salts
methane	CH_4	natural gas
polyethene	$(C_2H_4)_n$	polythene
potassium hydroxide	KOH	caustic potash
potassium nitrate	KNO_3	saltpetre
silicon(IV) oxide	SiO_2	sand, quartz
sodium carbonate	$Na_2CO_3 . 10H_2O$	washing soda
sodium chlorate(I)	$NaClO$	bleach (household)
sodium chloride	$NaCl$	common salt
sodium hydrogencarbonate	$NaHCO_3$	bicarbonate of soda
sodium hydroxide	$NaOH$	caustic soda
sodium silicate	Na_2SiO_3	glass
sodium stearate	$C_{17}H_{35}COONa$	soap
sodium thiosulphate(VI)	$Na_2S_2O_3 . 2H_2O$	hypo
sucrose	$C_{12}H_{22}O_{11}$	sugar
sulphuric acid	H_2SO_4	
zinc carbonate	$ZnCO_3$	calamine
zinc oxide	ZnO	
zinc sulphide	ZnS	blende

Sun, Moon, and Earth

figures shown are approximate

Diameter of the Earth	13 000 km	(7 900 miles)
Circumference of the Earth	40 000 km	(25 000 miles)
Distance from the Earth to the Sun	15 million km	(93 million miles)
Time for Earth to turn once	24 hours	
Time for Earth to travel once round Sun	365 days	
Distance from Earth to Moon	384 000 km	(240 000 miles)
Time for Moon to travel once round Earth	28 days	

Scientific units (SI)

length	metre	m
mass	kilogram	kg
time	second	s
speed		m/s
force	newton	N
weight	as for force	
work	joule	J
energy	joule	J
power	watt	W

length measurement

1 metre (m) = 1000 millimetres (mm)
= 100 centimetres (cm)
1000 metres = 1 kilometre (km)

mass measurement

1 gram (g) = 1000 milligrams (mg)
1000 grams = 1 kilogram (kg)
1000 kilograms = 1 tonne (t)

volume measurement

1 cubic centimetre (cm^3) = 1 millilitre (ml)
1000 cubic centimetres = 1 litre (l)
1000 litres = 1 cubic metre (m^3)

Useful conversions *(approximate)*

2½ centimetres	=	1 inch
8 kilometres	=	5 miles
5 kilograms	=	11 pounds (lb)
1 litre	=	1¾ pints
5 litres	=	1 gallon
1 tonne (t)	=	1 ton (Imperial)
3 metres	=	10 feet (ft)

Table of important elements

Element	Symbol	Proton Number	Element	Symbol	Proton Number
Aluminium	Al	13	Manganese	Mn	25
Argon	Ar	18	Mercury	Hg	80
Arsenic	As	33	Neon	Ne	10
Barium	Ba	56	Nickel	Ni	28
Bromine	Br	35	Nitrogen	N	7
Calcium	Ca	20	Oxygen	O	8
Carbon (graphite)	C	6	Phosphorus (white)	P	15
Chromium	Cr	24	Platinum	Pt	78
Chlorine	Cl	17	Potassium	K	19
Cobalt	Co	27	Silicon	Si	14
Copper	Cu	29	Silver	Ag	47
Fluorine	F	9	Sodium	Na	11
Gold	Au	79	Strontium	Sr	38
Helium	He	2	Sulphur	S	16
Hydrogen	H	1	Tin	Sn	50
Iodine	I	53	Titanium	Ti	22
Iron	Fe	26	Tungsten	W	74
Lead	Pb	82	Zinc	Zn	30
Magnesium	Mg	12	Uranium	U	92

Index